PRIVILEGED GOODS

GOODS

Commoditization
and Its Impact on
Environment and Society

PRIVILEGED GOODS

Commoditization and Its Impact on Environment and Society

by

Jack P. Manno

Executive Director
Great Lakes Research Consortium
State University of New York
Syracuse, New York

LEWIS PUBLISHERS
Boca Raton London New York Washington, D.C.

ISEE
International
Society for
Ecological
Economics

Library of Congress Cataloging-in-Publication Data

Manno, Jack
   Privileged goods : commoditization and its impact on environment and society / Jack P. Manno.
      p.   cm.
   Includes bibliographical references.
   ISBN 1-566-70390-5
   1. Industrial management--Environmental aspects. I. Title.
HD30.255 .M34 1999
658.4′08 21--dc21   99-043469
                                                                                              CIP

© 2000 by CRC Press LLC
Lewis Publishers is an imprint of CRC Press LLC

No claim to original U.S. Government works
International Standard Book Number 1-566-700390-5
Library of Congress Card Number 99-043469
Printed in the United States of America 1 2 3 4 5 6 7 8 9 0
Printed on acid-free paper

Contents

Acknowledgments

What I have enjoyed most about writing this book is that I haven't done it alone. I come to my subject as an independent, cross-disciplinary scholar, an activist, and a bit of a dreamer. Each of these roles is nourished by the connections I have with people in my life who challenge my thinking, fire my heart, and bless my soul.

My mother, Ginny, died during the time I was writing this book. Because my university had given me a leave of absence for research and writing, I was also able to spend the last 3 weeks of her life at her bedside with my father, Pete; my sisters, Eileen and Terri; my Aunt Phylis; and my wife, Cindy. The breaks from writing were spent reading to her, comforting her, playing cards, and doing crossword puzzles with my family. Breaks from her suffering were spent writing the words of this book. I hope something of Mom's interest in everyone and her zest for life made its way into the fabric of this book.

My own family has wholeheartedly supported my commitment of time and attention to this project. Cindy Squillace and Dianna and Daniel Manno are my grounding and my joy. My collective family of Dik Cool, Karen Mihalyi, Cora Cool-Mihalyi, Chris Perdue, Liz Cool, Derek Jutton, Mark Cool, and Robert Squillace, with their commitment to community and friendship, provide for me a model for an economy of caring and connection. I am incredibly wealthy in life-long friends, many of whom have listened to me at length while I sorted out the thoughts that fill this book. These especially include Gina Kellogg, Frank Ireland, Marc Servin, Roz Jacobs, Phil Rose, Beth Broadway, Melinda Wheeler, and David Marc. I also greatly appreciate the connections I have made and the support I have received through my involvement in Reevaluation Counseling.

Throughout the writing of this book I have had periodic "writing days" with my friend Mara Sapon-Shevin. While I was writing this book she completed hers, *Because We Can Change the World: A Practical Guide to Building Cooperative Inclusive Classroom Communities* (Allyn and Bacon 1999). Her commitment to helping students find their power to change their world was turned on me as well, and I came to believe in the importance of my project much more readily because of how much I believed in her project.

In the writing itself I have had a great deal of assistance. Chris Crysler organized much of the research and persistently kept me on track. Alex Antypas helped me with the writing itself, skillfully editing my drafts into clarity, and writing most of Chapter 6. Michael Connerton helped with the illustrations. Michael Mageau contributed his thoughts about the implications of ecological principles and ways to present some of the key concepts in this book. Luis Lemus Lopez assisted me in the early stages of this work. Lisa Porter, Joe Dadey, Jae Won Choi , Tom Ebert, and Nate Ermer, students at the College of Environmental Science and Forestry, have helped in the research. Several colleagues read portions of the manuscript and provided helpful comments: Mathis Wackernagel, Isidore Walliman, Lynne Woerhle, Tom Princen, Laura Westra, Orrie Loucks, and Ross Whaley.

Wendell Berry, although he doesn't know it, has been a major influence on this book through his collection of poems titled *The Timbered Choir*. While writing this book I took daily walks near my home and read from these poems aloud to myself. Nearly every day I'd read a passage that related uncannily well to an idea I was trying to present in prose. In this way Berry became a collaborator in this project. He gets much credit for inspiration and none of the blame for whatever interpretations I have made of his words. I highly recommend his book.

My colleagues at the College of Environmental Science and Forestry and the Great Lakes Research Consortium have been models of collegiality: supporting my work and challenging my thinking. Finally, for the past 2 years I have been President of Great Lakes United, a broad coalition of 170 organizations dedicated to the protection and restoration of the Great Lakes ecosystem. It is here that I have learned the practical challenges of trying to protect a particularly beautiful, important, and endangered part of our planet. Special thanks to Margaret Wooster, John Jackson, and the Board of Directors and staff of Great Lakes United.

About the Author

Jack P. Manno is a cross-disciplinary scholar and writer, the Executive Director of the New York Great Lakes Research Consortium, and an adjunct Associate Professor in the Faculty of Environmental Studies at the State University of New York College of Environmental Science and Forestry in Syracuse, NY. He is also President of Great Lakes United, a coalition of 170 organizations in Canada and the U.S. including environmental activists, First Nations and Native American organizations, conservationists, hunting and fishing clubs, and others working for the protection and restoration of the Great Lakes–St. Lawrence Ecosystem of North America. For the past 3 years Jack has been affiliated with the Global Ecological Integrity Project, a multi-disciplinary team of ecologists, philosophers, legal scholars, and economists working to improve understanding of the practical implications of the concept of ecological integrity in a range of ecosystems around the world. He has written extensively on the dynamics of social and political systems, including the militarization of the U.S. space program, the role of nongovernmental organizations in world environmental politics, and others. He can be contacted at jpmanno@mailbox.syr.edu.

chapter one

Introduction

This book addresses two questions: What are the most difficult obstacles in the way of effectively solving the environmental crises of our time? What can we do to overcome those obstacles? The answers I present here are a portion of a far more complicated story that will only be told in retrospect after we or our descendants have finally learned how to live lightly and well on the Earth.

Human attention is among the most powerful natural forces in the universe. When an individual decides to turn his or her attention to understanding something, accomplishing something, changing something, the resources available to that individual are mobilized, and the world begins to change. When groups of individuals who are organized in societies turn their collective attention to shared goals, the potential results are even more dramatic. Organized human societies have completely transformed the world. In many ways this book is about how this resource, human attention, is allocated. Since economics is fundamentally about the allocation of limited resources among competing uses, this book is about an economics of attention.

Humans have been remarkably successful, spreading throughout the planet, altering and shaping landscapes to improve conditions for our comfort and safety. The curiosity and creativity inherent in human nature has been unleashed in the production of multitudes of products and services that enhance life. The pace of this inventiveness is astounding. Those of us privileged with purchasing power and ready access to consumer goods can interact with an amazing system of production and distribution that makes available to us a garden of earthly delights unimaginable to all but a tiny elite of our ancestors. The earthly goods that surround our lives keep getting better and better. Our cars improve in speed, quiet, and comfort. Our ability to talk with each other anywhere, anytime, grows simpler, and the sound is clearer. The pharmacy shelves expand with new treatments for new ailments every day. The toothpaste dispenser gets better and easier to use. Music reproduction continues to improve in fidelity, as does ease of playback. Food arrives from all over the world, spicier and more interesting than ever. For every human need and desire there is a packaged, well-designed "something" to be purchased that will satisfy it. The global machine that produces the com-modities that make our lives richer, safer, and more comfortable is an amazing success, and many people everywhere in the world are understandably eager to join in the shopping spree. The basic political and economic objective of most modern nation-states has been to achieve increased prosperity for its people through economic growth, which in turn is pursued through improve-ments in economic efficiency (more output per dollar invested), technology, and trade.

What we have accomplished in the way of prosperity on Earth — and we have accomplished a great deal through this organized effort to promote economic growth — has required immense resources of human creativity, insight, and organizational capacity. We have had to discover the sources and invent the means for drawing from the natural world the energy and materials we need to produce our goods and services. Economic policy and practice and individual ambitions have been and are understandably coor-dinated toward this aim. The perspective of this book is that it is possible to achieve balanced, equitable, healthy, sustainable development, not only in the economy of commodities but also in the economy of care and connec-tion in all areas of human life. For all the success of economic development, huge portions of human potential are left deliberately underdeveloped as a result of a distorted allocation principle guiding the allocation of human attention. We need to begin to address this imbalance of development, because in this imbalance lies the source of many of the crises that now threaten the health of the planet.

Human beings live in groups.[1] We cooperate with each other to enhance our ability to achieve individual material well-being and solve the four basic challenges of economic development: obtaining raw materials, transforming those raw materials into things we can use, sharing the efforts and distrib-uting the products in a way that keeps the community together, and storing (or insuring) what we need for later periods of scarcity. These challenges have been met in countless diverse ways by the multitude of cultures that have existed on Earth. To be sustainable, a society must not only solve these basic challenges but must do so in an ever-changing environment. These changes include the way the environment is affected by human activities. In order to thrive, human communities must be flexible and creative. Infor-mation about the world has to feed back into the development process and adjustments then have to be made.

While we have been very good at building, destroying, and building anew, we have been very bad at conserving, preserving, and sustaining. The great challenge of our times is to overcome this imbalance. In the future, if we are to have a future worth looking forward to, we will have to become as good at protection as we are at innovation, as good at conservation as we are at transformation, as good at preservation as we are at creation, and as good at sustaining as we are at altering. "May you live in interesting times!" So goes the saying, thought to be ancient and Chinese, which may be a blessing or a curse depending on one's perspective. We who are alive, reading and writing at the beginning of the twenty-first century certainly live in interesting and challenging times.

The fascinating and tragic dilemma now challenging us is the fact that, while the amazing abundance of people and goods is being produced, the ecological health, stability, and integrity of the Earth itself are being threatened by the byproducts and side effects of the very patterns of economic behavior that have created this abundance. The economy that has been so good for many is now threatening to become very destructive for all. According to the United Nations Human Development Report 1998:

> Runaway growth in consumption in the past 50 years is putting strains on the environment never before seen.

- The burning of fossil fuels has almost quintupled since 1950.
- The consumption of fresh water has almost doubled since 1960.
- The marine catch has increased fourfold.
- Wood consumption, both for industry and for household fuel, is now 40% higher than it was 25 years ago.[2]

These figures must be considered in light of who is doing the consuming. Consider the following facts, also from the UN's Human Development Report:

> Inequalities in consumption are stark. Globally, the 20% of the world's people in the highest-income countries account for 86% of total private consumption expenditures — the poorest 20% a miniscule 1.3%. More specifically, the richest fifth:

- Consume 45% of all meat and fish, the poorest fifth 5%.
- Consume 58% of total energy, the poorest fifth less than 4%.
- Have 74% of all telephone lines, the poorest fifth 1.5%
- Consume 84% of all paper, the poorest fifth 1.1%.
- Own 87% of the world's vehicle fleet, the poorest fifth less than 1%.[3]

Given a choice, the poorest fifth of world's population would probably like to live like the wealthiest fifth. The impact of that level of consumption on the Earth's environment and resources would be staggering. But such levels of inequality — and they are growing — are not politically sustainable forever. Either consumption will increase dramatically among the presently poor, or the world will descend into bitter conflict over the distribution of wealth and resources, or we will advance to a new economy of equity and sustainability. The next great stage of human history will be dominated by our struggle to resolve the tragic dilemma of the modern economy. Can we figure out how to build communities and individual lives that thrive and prosper without damaging the ecological life support systems upon which all health and prosperity ultimately depends? Is it possible for large numbers of human beings, perhaps all human beings, to live prosperous lives without destroying the Earth? Can we design and carry out such an economy? Understandably, we want more than simply to survive, we want to prosper. By prosperity I mean having the material well-being to meet basic human needs and rational aspirations for comfort and security. Moreover, prosperity should not be a privilege for the lucky or powerful few but a reasonable goal for vast numbers of people worldwide, ideally for all.

The most critical threat to continued human development and improvement is the fact that the scale of human use and waste of material and energy resources is threatening to disrupt the earth's life support systems, making the natural environment more unstable and less hospitable to the forms of life, including human, that have developed under the Earth's present conditions.[4] Mathis Wackernagel and William Rees have developed a measure of the amount of land space required to support a given level of resource use, the so-called "ecological footprint." Wackernagel has observed that to support the world's population at the standard of living enjoyed by the industrialized wealthy nations of the North would require a land base equivalent to three additional Earths.[5] If we continue down this path, life is bound to get harder for many, if not most, of the world's people, and for far too many of them life is already barely endurable. Even if we learn to adapt to decreases in the *per capita* availability of productive agricultural lands, ocean fisheries, clean water and air, and other resources on which our lives depend, the pressures on many, perhaps most, of the Earth's creatures may well prove to be too much for them to survive. We have a moral obligation not to let that happen. What is obvious to most anyone who takes the time to really look at the environmental trends at the beginning of the twenty-first century is this: if we don't change direction we'll soon end up where we're heading. It's not a pretty picture.

What we must do now seems obvious: do more with less. In every sector of the economy we have to dramatically reduce the amount of energy and material we use to satisfy human needs and wants. We must first redefine what we mean by efficiency, by changing it from its present meaning of maximum output per dollar invested, to something like maximum service per unit of energy and material produced and consumed. We must then set all our scientific creativity and intelligence to the task of making our patterns of production and consumption increasingly efficient in this way. This is a great undertaking that could and should occupy all people. There are many ways to accomplish this type of efficiency. It could be a renaissance of huge proportions as we make all aspects of life more efficient in these terms. There would be a revival of skills and knowledge as we grow more food with less fuel, work *with* rather than *against* nature's cycles and flows, use less energy for everything, repair more than we replace, share more than we consume, rely on each other more, and use more creativity and less material and energy. In order to accomplish this we need to change how we allocate our attention. Instead of focusing on goods and services (commercial commodities) as ends in themselves, we must focus on the *service* such goods and services are meant to provide. There may be countless ways to satisfy any given need or solve any problem. Some of them may be commercially viable solutions and others may not be. Some of the options that may best solve any problem may be more environmentally benign or more socially beneficial than others. When we allocate our attention predominantly toward commercially viable options we abandon our ability to allocate our attention according to other goals. We must change the primary allocation principles for human attention so that reducing human impact on the planet and building communities that thrive are the primary goals.

This is a technological, social, and political challenge that can and should engage and excite everyone. It is not just an engineering and scientific problem, but one which will call upon the creative intelligence of community activists, educators, gardeners, launderers, cooks, childcare providers, everyone to be solved. As a result of this creative response to the challenge of thrift, life will be better in many ways. Such a vision is the next stage of economic development. The path to it has been laid out by countless ecologically inspired optimists working at transforming every aspect of life — transportation, housing, neighborhoods, agriculture, health care, education, and on and on. Some have called it Sustainable Development, to contrast it with the patterns of development currently dominant in most of the world, and rapidly spreading, which continue to use and waste greater and greater quantities of energy and materials and which are clearly unsustainable as a result.

So why, when we understand some of the consequences of reckless economic development and the need to be careful about our impacts on the natural world, and having understood this for decades and perhaps for thousands of years, is it so profoundly difficult for us to do it? Why, when it is so clearly rational and correct to do so, can't we put at least as much attention and resources toward conserving energy and materials as we do toward mining and harvesting more and more? Why can't we put as much attention and resources on developing good public transportation as we do on more muscle-bound jeeps? Why don't we reduce waste generation before we reuse, and reuse before we recycle, and recycle before we burn or landfill? Why not do as much research into organic agriculture as the fertilizer and pesticide industries do on their R&D? Why not spend as much on disease prevention as we do on pharmaceuticals and high-tech treatments? Why not spend more on energy conservation than oil exploration? Given the threats to the habitability of our planet and the mess we may be leaving to our children and grandchildren, the choices are logical and obvious. And yet, there is little sign that in transportation, land use, agriculture, health care, or even environmental protection we will soon fund the development of ecologically preferable options anywhere near as generously as we have ecologically destructive ones. If anything, the opposite is true; ecologically destructive ways of living are continually spreading into societies and cultures that once managed to live more frugally and in balance with nature. Why?

The first step is to get clearer about what the problem is and what it is not. There are many suggestions as to the root cause of unsustainability: human greed,[6] ignorance,[7] spiritual detachment and/or malaise,[8] addictions,[9] the flawed logic and practice of contemporary economics,[10] the ideology of competitive growth under capitalism,[11] or simply the inevitable consequence of human activity.[12] This book presents the case that the problem is not simply about individual human greed and fear of scarcity, but also about the *systems* of rewards in place that reinforce greed and cater to fear. It is not only about ignorance of ecology or the problems of environmental degradation, but also about *self-reinforcing processes* that reward environmentally destructive choices. It is not just about spiritual malaise or addictions to material goods, but about political decisions that make self-reliance and mutual aid increasingly difficult to choose. Unsustainability is, in the end, the inevitable consequence of particular dynamics in the modern political economy that systematically favor exactly those forms of goods and services that have the greatest environmental impact.

If we can identify the most significant obstacles to building ecologically informed, sustainable societies, then we can also determine how to overcome those obstacles. Although many can be identified — lack of ecological knowledge and understanding, lack of political will, a mindset locked into the ideology of perpetual growth, extreme individualism, a spirituality devoid of ecological consciousness, severe population pressures, the desire of the poor to live like the rich, lack of resources, inertia, and many others — the most important by far are the incentives and disincentives that lie at the core of modern economies.

The danger of any criticism — certainly this book criticizes the patterns of modern economic development and their effects on the Earth's health and ecological integrity — is to cast the subject into contrasting sides of good and bad. Throughout this book I will remind the reader that the problem of unsustainable forms of economic development is not inherent in markets or economics *per se*, but is rather the distortions in economic and social development that can occur when either self-regulating market systems or self-enriching state power predominate in the choices made about the allocation of human attention and other resources. The problem is not the wealth-producing economy, but the distorted ways wealth is distributed. The problem is not technology, but the distortions caused when most technology is directed toward further development of a narrow range of particular kinds of products. The problem stems from the ways human attention, creativity, and intelligence are harnessed, like our technology, toward a narrow range of human possibilities. Despite the problems addressed in this book, humans have accomplished amazing things through our present stage and systems of economic development.

In 1992 I attended the United Nations Conference on Environment and Development in Rio de Janeiro, Brazil, as part of a small group of scholars who were studying the role of nongovernmental organizations in world environmental politics.[13] What struck me most after listening to the many days of deliberation was how much consensus actually existed and how well-informed the conference delegates were. In addition, there were nearly 20,000 environmental activists from all over the world sharing information and arguing strategies in the parallel NGO Summit in Flamingo Park. Despite some reluctance on the part of the U.S. delegation, representing the opinions of the conservative administration of President George Bush, there was very little actual disagreement about the seriousness of threats to the habitability of the Earth and the danger this posed to the well-being of humanity. There was also surprisingly little disagreement about what solutions were needed. In the months leading up to the Earth Summit, thousands of "experts" arrived at a consensus document known as *Agenda 21*,[14] which in 40 chapters laid out a consensus plan for achieving sustainable development in the twenty-first century. *Agenda 21* was neither radical nor new, because what is needed to reform economic development and make it more sustainable has been known for a long time. Prior to the Earth Summit, the World Commission on Environment and Development — known as the Brundtland Commission after its chair, Gro Harlem Bruntland, the former Prime Minister of Norway — had issued its report[15] concluding that the rich with their profligate waste of materials and energy and the poor with their desperate overharvesting of local resources for survival and their inadequate capacity to manage domestic wastes are both responsible for the global environmental plight, and that the rich clearly had the greatest burden of responsibility with their access to the greatest share of the world's wealth and resources. The Bruntland Commission and *Agenda 21* both reported that the wealthy minority of the Earth's people, often referred to simply as the North, and the poor majority, the South,[16] had what the other needed for sustainable development: the poor of the South often developed ingenious systems of mutual aid, recycling, and thrift from which the North could learn much, while the North had created new energy and materials-conserving technologies as well as the know-how to transfer the technology of environmental protection to the crowded cities of the South, which desperately need them. Major new investments would be required in what has come to be known as "appropriate technology," "ecological design and engineering," and "ecosystem management of natural resources." These technologies and management practices are characterized by the integration of social and economic systems with the knowledge of ecological processes and their implications. The technology and know-how for designing and implementing ecologically informed systems of production, energy transmission and use, transportation, construction, and so on are presently available to experiment with and implement widely.

There have been many efforts to summarize what it means to live, work, and produce in an ecologically appropriate manner based on an understanding of fundamental ecological principles. These will be discussed in some detail in Chapter 7. At some level the prescription for sustainable living is simple. Nature exists in a multifaceted and dynamic balance between organisms and their environments, and we need to maintain that balance as best we can. Too much of almost anything is never a good thing: it leads to imbalance and simplification. There are countless excellent expressions of fundamental ecological principles. The ancient hunters and gatherers learned them from observation and experience, the first farmers drew from them, the Buddha followed them, Christ preached them, Thoreau went to Walden Pond and rediscovered them. Today they are part of the popular wisdom: In simplicity one is close to the sacred. Everything is connected. There's no such thing as a free lunch. In nature all waste is food. Diversity of approaches improves the chances that at least one will succeed. Matter and energy can neither be created nor destroyed. All order is bought at the cost of disorder somewhere. These and many similar observations of the way the world works are generally accepted, and equally generally ignored.

There is no shortage of good information. There are millions of excellent books, articles, and pamphlets on alternative agriculture, appropriate technology, renewable energy, soil conservation, sustainable forestry, industrial ecology, and other steps to be taken to create an ecologically sound and sustainable economy. Not only are there good ideas but thousands of good experiments and brilliant tinkerers who are designing and implementing creative solutions in all areas of life; whole communities and ecovillages are popping up and learning how to live well while living lightly on the Earth. And yet, despite this knowledge and available information, very little progress has been made to slow, never mind reverse, the planet-wide environmental devastation caused by human failure to develop a sustainable way of life on a large scale.

The literature and know-how of the alternative and appropriate technology movements, organic agriculture, energy efficiency and many others goes back at least to the early 1960s.[17] But despite the fact that we have known for many years what is needed, we have made remarkably little progress in implementing sustainable development worldwide. In fact, all the trends since 1992 are moving in exactly the opposite direction. During the Earth Summit the *Los Angeles Times* reported it this way:

> World leaders will gather for a global environmental summit in Rio de Janeiro on June 3. If it's a typical day…

- 250,000 people will be added to the world's population.
- Up to 140 species of living creatures will be doomed to extinction.
- Nearly 140,000 new cars, trucks and buses will join 500 million already on the road.
- Forest covering an area more than one-third the size of Los Angeles will be destroyed.
- More than 12,000 barrels of crude oil will be spilled into the world's oceans.[18]

The momentum of growth seems almost unstoppable. Lester Brown and colleagues at the Worldwatch Institute summarize the present growth scenario best:

> Global economic output, the total of all goods and services produced, grew from $5 trillion in 1950 to $29 trillion in 1997, expanding more than twice as fast as population. This increase of nearly six fold boosted incomes rather substantially for most of humanity. Growth of the world economy from 1990 to 1997 exceeded the growth during the 10,000 years from the beginning of agriculture until 1950. Economic output per person climbed from just over $1,900 in 1950 to nearly $5,000 in 1997, a gain of 163%.... At the same time the use of lumber has more than doubled. That of paper increased six fold, the fish catch increased nearly five fold, grain consumption nearly tripled, fossil fuel burning nearly quadrupled, and air and water pollution multiplied several fold.... If the economy was to expand only enough to cover population growth until 2050, it would need to grow from the $29 trillion of 1997 to $47 trillion. This, of course, would merely maintain current incomes, unacceptable though they are for much of humanity. If, on the other hand, the global economy were to continue to expand at 3 per cent per year, the global output would reach $138 trillion in the 2050.... If the world cannot simultaneously convert the economy to one that is environmentally sustainable and move to lower population [growth] ... economic decline (from depleted resources) would be hard to avoid.[19]

One way to avoid this economic and environmental decline and convert to an environmentally sustainable society is to weaken the relationship between human welfare and excessive material and energy consumption, to achieve a high quality of life with a minimum of stuff. The sad fact is, we are going in the opposite direction. Why? Why is it that when we know what we need to do to protect and restore the health and ecological integrity of the planet we have made so little progress in doing so? That is the question that has inspired this book.

The critical questions are, can we make the necessary adjustments in our priorities and reform our environmentally and socially destructive economic practices in time? Can we live well while also living lightly upon the Earth? Can we live simply so others can simply live? In this book I argue that the propensity toward destruction, the tragic flaw of the modern economy, rather than being an inevitable outcome of human settlement,[21] can instead be understood as a result of certain largely hidden and misunderstood economic forces that drive social and economic development toward ever increasing mobilization[22] of energy and materials. By taking a close look at what I call "commoditization,"[23] this book helps explain how the economy organized by humans to serve individual aspirations for material well-being has ended up, through the unintended consequences of certain system features, threatening the very life support systems upon which all human well-being depends.

Before proceeding with the argument, first let me share a little personal history. On August 20, 1968 the Soviet Army invaded what was then Czechoslovakia. I remember that day well, because it happened to be the day I had been invited to the United Nations by the British ambassador at the time, Hugh Lord Carrington. I was 17. Earlier that year, Lord Carrington had been the guest speaker at our regional High School Model United Nations in the Capital District of New York State. Having been intrigued by global affairs since I was in seventh grade, I had recently been elected President of the new session of the Model UN. Before he left, Lord Carrington invited the outgoing and incoming presidents, my friend Roger and I, to visit the United Nations and join him for a typical day as a UN ambassador. As it turns out the scheduled day was anything but typical. When we arrived at the UN headquarters, the place was astir with preparations for an emergency session of the Security Council for that evening. Ambassador Carrington sent us his regrets that we would not be able to spend the full day at his side as intended, but said that he would still join us for lunch and that later that evening we would be allowed as his guests in the Security Council chambers as the Council debated a response to the Red Army's invasion. At lunch the Ambassador seemed to be in a pensive and sad mood. After some small talk, he spoke quietly and at length, looking not at us but out the window and across the East River. He was discouraged and not looking forward to the evening's debates, which everyone knew would be futile with the Soviet Ambassador present and prepared to use his veto power to thwart any attempt to express UN support for the beleaguered Czech rebels. He talked about the dangers that lay ahead for our generation. He wanted us to help solve what he saw as the great mystery of human development: why is humanity so good at building great monuments, filling the Earth with trinkets and toys, and inventing new and more powerful means of destruction while at the same time so bad at learning how to live well and peacefully with each other? What was wrong with human intelligence that it was so out of balance, that we showed such creativity and organizational genius when it came to material things and such ignorance when it came to interpersonal and intercommunity relationships? He asked us to devote ourselves to helping human ethical and spiritual development catch up with human technological development. I thought about his request for many years, and this book is part of my response.

The following year U.S. astronauts walked on the moon. The feat, completed less than a decade after President John Kennedy committed the U.S.

space program to doing it, was an amazing example of what can be accomplished when scientific knowledge, engineering skills, attention, money, organization, and will are harnessed toward a goal. Why, many a critic pleaded, can't we turn such resources and attention onto the problems of poverty, environmental degradation, war, and racism here on Earth? President Richard Nixon responded by appointing former NASA leaders to head urban development and anti-poverty organizations in the U.S., with less than spectacular results. What was it about the problems of spaceflight that were amenable to technological solution, while the social problems haunting the planet's surface seemed at times only to worsen with advances in technology. This seemed to me at the time another example of Carrington's problem: that technological advances continued to race ahead of advances in human relationships with each other and the Earth.

Today the gap between technological and social development continues to trouble us as we face the challenge of sustainable development. What is in the way of turning at least as much of our collective human energy and creativity now toward protecting and sustaining our air, water, soils, and communities as we have in the past into producing cars, computers, gadgets, new fabrics, new toothpaste, better cosmetics, improved drugs, etc.? The question has many corollaries, such as why do we put so much more of our health resources toward treatment of illness than toward maintaining health? Why do we invest so much more in soil additives than we do we in soil protection? Why do highways get so much more resources than urban public transportation? Why does the design of weapons of war get so much more research and development attention than the development and promotion of the skills and techniques of peacemaking? Why does it cost so much more to repair a radio than it does to purchase a new one?

Part of the answer to each of these questions has something to do with the fact that it's simpler to work with things and engineering abstractions than it is to work with people and systems. It's easier to pay attention to commercial goods than the services they are meant to provide. It's easier to produce one huge power plant and distribute the electricity than it is to retrofit homes and redesign manufacturing to save the same amount of electricity the plant can produce. It's easier to produce medicines and devices than it is to educate people in healthy ways of life, to coordinate public hygiene programs, or to create the conditions in which people care enough about themselves to take good care of their bodies. It's easier to threaten with the newest and best weapons than it is to resolve conflicts. It's easier for everyone to get into their own cars on their own schedule than it is to coordinate the movements of large numbers of people in mass transit systems. It's easier to design spacecraft and calculate launch trajectories than it is to build healthy communities. It's easier to throw things out and replace them with new things than to organize a system of repair and maintenance. At one level, the source of the imbalance between technological and social development is just that *things* are easier to work with than *people*. We naturally settle for that which is simplest.

Although the propensity to favor ease and simplicity has some power to explain Carrington's dilemma, there is much more to it than that. After all, the cars, the weapons, the spacecraft, the medicines, and all the goods of a modern economy require for their production and distribution the coordination of the labor and creativity of thousands of individuals and millions of component activities. They require the mobilization of vast amounts of energy and materials and the investment of large quantities of capital, wealth produced either through previous productive activities or through theft, looting, or exploitation; usually a combination of each. There is something else that each of the examples of Carrington's dilemma demonstrates. The medicines and equipment of treatment are somewhat easier to own, package, and sell than are the tools and techniques of prevention. The power plant not only produces power, but also concentrates it. The weapons of war are simpler to sell than the skills of negotiation. Likewise, a transportation system built around personal automobiles involves many more opportunities to package and sell than does a coordinated system of public transit. Soil additives, chemical fertilizers, and insecticides are all products patented, packaged, distributed, and sold. The farmer who knows her soil and through skill and hard work recycles nutrients from the farm and protects the soil from erosion and overuse has as her most important product her knowledge and skill, which cannot easily be packaged and sold. The new radio is easier to market than the skills of the repairman.

One more thing is true about these examples. Each involves the delivery of a service: transportation, nutrition, peace and security, health, music. What we see is that for every human need there is a tendency to satisfy it with those commercial goods that can be most easily bought and sold. This is the theory of *commoditization* that is the subject of this book. The incentives and disincentives that structure the economy drive people to preferentially harness the resources of society and nature for transformation into commodities for sale in markets. As we will see, many of the qualities that characterize commodities are precisely those qualities that concentrate and mobilize energy and materials. In the end, commoditized satisfactions for human needs and desires outcompete alternatives that would require less energy and materials. In effect, commoditization acts like a selection pressure, choosing certain types of goods and services over others. This process of selection is a function of the laws, institutions, and cultural practices that we have put into place to maintain a growth economy — a system of resource exploitation, transformation, and consumption that is predicated on indefinite expansion.

In the end, Carrington's dilemma is misstated. The problem is not only about overdeveloped technology and underdeveloped human spirit. Technology itself is not overdeveloped but rather it is malformed, maldeveloped by excessive emphasis on one kind of technology — technology at the service of the production of commercial goods, commodities. Science and technology that has the potential to contribute toward the mass production of commercial goods receives by far the lion's share of research and development resources. What we think of as "appropriate technology" or "alternative technology" tends to be designed and developed for specific tasks in a particular location, as part of a particular system and is far less amenable to commercialization, mass production, and distribution. The problem then is not technology or the technological imagination but rather the distorted ways in which the technological imagination is directed in a commoditized economy.

The more a society and an economy is organized for the purpose of producing goods and services for sale, the more powerful is the force of commoditization. Over time, more and more of life's needs are satisfied by commodities. In general, it is easier to claim ownership over, package, market, transport, and sell commodities than it is to own, package, market, and sell people's skilled labors, and their knowledge of particular people and places. Thus the more society holds the highest privileged place for goods and services that can be marketed and sold, the more human progress and development becomes equated with the production of more and more commodities. This is inevitable in a society that gives highest priority to economic growth over all other measures of quality of life. The only things that count in measures of economic growth are by definition those goods and services that are bought and sold. The more successful a society is in creating new human needs and matching them with commoditized need satisfactions, the more successful a society is in stimulating and maintaining economic growth.

At its core, commoditization is the continuous pressure to transform as much of the necessities and pleasures of life as possible into commercial commodities. Those things that are difficult to commercialize are systematically denied the kind of research and development attention that goes into commodities. In Chapter 2 we will describe the characteristics that makes something suitable for sale, that makes something commoditizable. Those things that are difficult to commercialize are usually so because they involve specific relationships of caring and mutual aid between people or relations of connection between people and a particular place, culture, or ecosystem. This will be described in much more detail in the following chapters. For now, suffice it to say that increasingly, commercial commodities have replaced other forms of need satisfaction in the lives of most people. The more this happens, the more of the Earth's resources are turned into commodities, and the more the waste products of consumption are pumped back into the biosphere. The logic of decommoditization suggests that the more we develop our relationships of mutual aid and connection with the Earth, the less we need to rely on purchased inputs to make our lives go well, to prosper. This book argues that the process of commoditization is historically contingent and maintained by institutions and policies that can, if we desire it, be changed. It is possible to give as much attention in terms of research and development resources to the economy of care and connection as we do to the commercial economy of commodities. As commoditization shapes the direction and evolution of economic development, the result is the increasing use and waste of the Earth's natural resources with all its accompanying ecological consequences. Chapters 2 and 3 describe commoditization and the way it has affected the economic evolution of modern societies.

The pressures of commoditization, I will argue in this book, also result in a deep and relentless form of oppression. The notion of oppression carries with it both the idea of political and social injustice as well as the sense of people's creativity and spirituality being weighted down and suppressed by forces outside the control of oneself. In one sense we are all oppressed by the way these economic forces make it exceptionally difficult to use our best thinking and most creative skills to build the future we want and need. By systematically favoring highly commoditized goods and services over less commoditized alternatives, our best and most creative minds and our most enthusiastic young people face dashed hopes, limited potential, less than full reign for their creative development. Much of what makes social life successful resists commoditization, despite, as we shall see, heavy pressures to commoditize what is still outside the formal economy: the relationships of care and connection between people, and between people and the Earth. Commoditization renders these aspects more and more peripheral to the main stem where society's resources flow. As a result, the people who provide these types of goods and services, who participate in the economy of care, grow relatively more impoverished. Chapter 5 presents this understanding of oppression and its effects in more detail.

Chapter 6 traces the history of public policies as they have both encouraged commoditization and tried to deal with its consequences. The lesson here is that, myths of the Invisible Hand coordinating economic development aside, the guiding hand of powerful interests have been quite visible and deliberate. Public policies have exacerbated the negative effects of commoditization and made it more difficult to empower its victims to resist. While commoditization as an economic selection pressure and organizing force may begin as an unintentional consequence of human's well-intentioned efforts to develop and make a better life, it is not inevitable that it should be allowed to destroy the environment and impoverish public life. It has always been possible to implement public policies that provide countervailing forces to commoditization that support the protection and development of the economy of care. Much of the history of economic policy-making in industrial societies has been about trying to locate the balance between encouraging commoditization as essential for economic growth and prosperity and recognizing that certain sectors of society and certain aspects of human life benefit most from commoditization while others benefit least. It is now possible to develop public policies that optimize the benefits of both and find a balanced, ecologically sound approach to economic and social development.

Once commoditization is fully understood in both its positive and negative aspects, it becomes possible to imagine and then design and implement public policies that encourage a balance. A vast array of approaches to environmental protection have been suggested and many tried. Chapter 7 considers many of these as to how well each is likely to provide a counter-balance to commoditization. It describes policies designed to encourage investment in sectors of the economy most resistant to commoditization: subsidy reforms, investments in energy and materials efficiency, and product and service substitution. It also discusses tax reforms and regulatory reforms as reasonable options for creating countervailing pressures to commoditization. By focusing on commoditization, it becomes possible to evaluate the potential of different policies to effectively nurture the reinvention and renaissance implicit in the vision of sustainable development. The final chapter also describes the potential for building social movements worldwide to bring about the new economies of ecological balance and design. The resolution of the global ecological crises and the unleashing of the renaissance of economic, technical, and social creativity it makes possible require that we understand commoditization and develop the policies and programs needed to overcome its effects.

There is much confusion among environmentalists and others about what is "natural" and what is not. It's "natural" for humans to transform our environment to suit our ends. It's in the "nature" of human creativity to come up against the threats and limits imposed by "natural" conditions such as gravity, the presence of disease microorganisms, floods, and many others, and then to invent methods and tools to protect ourselves from the threats in our environment and to find a way around, through, or over the obstacles our environment places in our path. We do that by cooperating with each other, taking care of each other, paying close attention to and studying the natural world, and using science and technology to increase our abilities and capacities. Lord Carrington was wrong about the problem being flawed human nature that does well with technology but badly with relationships. The flaw is in the distorted, confused view of the world caused by misguided economic development in which commercial commodities are systematically privileged over relationships of care and connection. By paying attention to the way things work, human intelligence comes up with unique and creative solutions to every problem. Today one of the most crucial limits to humans living well together is the way the byproducts of the global economy threaten the habitability and hospitality of the Earth. To respond to the challenges before us, we need to pay attention to the way the Earth's ecological systems react to the changes wrought by the expanding human economy. It is entirely within the capacity of human intelligence to reform the human economy so that it has less of an impact on the Earth's life support systems. It is to that end that this book is directed.

It should be obvious by now that the subject of this book is huge. Within the limits of my own ability to understand how economic patterns have environmental consequences, I am offering my thoughts about the phenomenon of commoditization. I think it is a useful construct. Because of the scope of the subject, much of the discussion will be general and often abstract. There is not nearly enough hard evidence to prove a theory of commoditization. No one to my knowledge has ever broken down economic data into categories of high, middle, and low commodity potential. Such data would be needed to definitively show how the economy distorts development in the ways I suggest. I hope the reader will take this book as I intend it to be used — as a contribution to thinking about patterns of economic development and what needs to happen to make our future economy equitable, sustainable, and supportive of the people who do the important work of care and connection.

Notes

1. Carrithers, M., *Why Humans Have Cultures: Explaining Anthropology and Human Diversity*, Oxford University Press, New York, 1992.
2. United Nations Development Program, *United Nations Human Development Report 1998*, Oxford University Press, New York, 1998, p. 2.
3. United Nations Development Program (1998).
4. I am not going to recite here the litany of environmental problems associated with the impacts of human economic activity. There is a large and readily available literature on this. I recommend the State of the World Reports published by the Worldwatch Institute. A good summary of the major global impacts can be found in Vitousek, P.M., Mooney, H., Lubchenko, J. and Melillo, J., Human domination of Earth's ecosystems, *Science*, 277(25), 494, July 1997.
5. Wackernagel, M., Onisto, L., Callejas Linares, A., Lopez Falfan, I.S., Mendez Garcia, J., Suarez Guerrero, A.I., and Suarez Guerrero, Ma., G., *Ecological Footprints of Nations, Report of the Centre for Sustainability Studies*, Universidad Anahuac de Xalapa, Mexico and The Earth Council, Costa Rica, 1997.
6. Hardin, G., The tragedy of the commons, *Science*, 162, 1243–1248, 1968.
7. Orr, D.W., *Ecological Literacy: Education and the Transition to a Postmodern World*, State University of New York Press, Ithaca, NY, 1992.
8. Gore, A., *Earth in Balance: Ecology and the Human Spirit*, Houghton Mifflin, Boston, 1992.
9. Catton, W.R., *Overshoot*, University of Illinois Press, Urbana, IL, 1980.
10. Hall, C.A.S., Economic development or developing economics: What are our priorities? In *Ecosystem Rehabilitation*, Wali, N.K. (Ed.), SPB Academic Publishing, The Hague, the Netherlands, pp. 101–126. See also Daly, H.E. and Cobb, J.B., Jr., *For the Common Good: Redirecting the Economy Toward Community, the Environment and a Sustainable Future*, Beacon Press, Boston, 1989.
11. Herman Daly (1991) has stated this eloquently in a parable. "As long as we remain trapped by the ideology of competitive growth, there is no solution. We are reminded of the South Indian monkey trap, in which a hollowed-out coconut is fastened to a stake by a chain and filled with rice. There is a hole in the coconut just large enough for the monkey to put his extended hand through but not large enough to withdraw his fist full of rice. The monkey is trapped only by his inability to reorder his values, to recognize that freedom is worth more than a handful of rice."
12. Cross, J.G. and Guyer, M.J., *Social Traps*, University of Michigan Press, Ann Arbor, MI, 1980. See also Rees, W. E., Patch disturbance, dissipative structure and ecological integrity: a "second law" synthesis, In *Global Ecological Integrity*, Pimentel, D. (Ed.), Island Press, Washington, D.C., in press.

13. Princen, T. and Finger, M. (w/contributions of Manno, J.P. and Clark, M.L.), *Environmental NGOs in the World Politics: Linking the Local and the Global*, Routledge, London, 1994.

14. http://www.igc.apc.org/habitat/agenda21, 7/29/99.

15. World Commission on Environment and Development, *Our Common Future*, Oxford University Press, New York, 1987.

16. For the purposes of this book, I will use the inadequate terms "North" and "South" because they are the least loaded of all the available terms. There is clearly some need to distinguish in our discussion between the minority of the world's people who live in the industrialized, relatively wealthy, and highly developed nations, which tend to be in the Northern temperate zone, and the majority who live in the less industrialized, relatively poor, and less developed nations, which tend to be nearer to the equatorial zones. There are no good ways of referring to these blocks of humanity. The global spread and integration of the industrial economy means that there are large pockets of the industrialized, commercialized "North" in the urban areas of virtually every southern country. Likewise, there are large pockets of poor people living largely outside the organized economy in rural areas and declining city centers in many of the richest countries of the North. Another common shorthand contrivance is to refer to the First, Second, and Third Worlds: the First referring to the major industrialized nations; the Second to the countries of the former Socialist Bloc of the Soviet Union and Eastern Europe; and the Third being the countries of the rest of the world striving to achieve economic development. This categorization scheme is probably useless with the dissolution of the former Soviet Union and Eastern Bloc. Other categories that are sometimes used include industrialized, newly industrialized, and emerging economies; developed, developing, underdeveloped, etc. All of these carry implications that the patterns of economic development in the industrialized North are the models toward which the whole world should strive. The problem is that the level of material and energy use in the industrialized North cannot be copied in the rest of the world without placing unacceptable burdens on the Earth's life support systems. So in terms of sustainable development, the entire world, both rich and poor, is underdeveloped or maldeveloped, each in its own way.

17. See Dickson, D., *Alternative Technology and the Politics of Technical Change*, Wm. Collins & Sons, Glasgow, 1974.

18. Dolan, M. and Abramson, R., Expectations for summit come down to earth, *Los Angeles Times*, home edition, Part A, June 1, 1992.

19. Brown, L.R., Gardner, G., and Halweil, B., *Beyond Malthus: Sixteen Dimensions of the Population Problem, Worldwatch Paper 143*, Worldwatch Institute, Washington, 1998, pp. 58–60.

20. United Nations Development Program (1998), p. 37.

21. Rees, W.E., "Patch disturbance" (in press). William Rees argues that the eco-logical characteristics of human beings as large animals who live in groups means that even in a hypothetically low-impact society humans would in-variably reduce the biodiversity and ecological integrity of their natural sur-roundings in obtaining food and the materials of daily living. This is most likely true but underestimates, in my opinion, the capacity of human beings to understand our inherent impact on the natural world and design methods of production and organization that minimize and compensate for this ten-dency — a capacity that is presently limited by the constraints described in this book.

22. I use the term *mobilization* rather than the more frequently used *consumption* because *consumption* implies that something is consumed and is gone, whereas the material and energy involved in producing our goods and services are often removed from where they are stored in Nature and are *mobilized,* caused to move from storage into circulation temporarily as goods and services and then as wasteproducts of one form or another.

23. I use the term *commoditization* rather than the more commonly used *commod-ification* to emphasize the active nature of the process being described. It carries more of the sense of an active verb, *to commoditize,* rather than the passive, *to be commodified.* The only other use of the term commoditization I am aware of is in Ayres (1998). There the term is used to describe changes in the organization of production in which the design, manufacturing, packag-ing, and distribution of a product is increasingly segmented where those with the least potential value added are concentrated in the South while the activ-ities with the most value added remain in the North. As a result, investments in manufacturing in the South often take on the character of investment in primary commodities and as such are subject to the same problems of falling prices and export earnings that have hampered Southern exporters of primary commodities. Very little in terms of skills or technology transfer is gained by the host country when these sorts of investments are made. This meaning of commoditization is very similar to the one used in this book, in that it is the result of investor preference for mobility, packageability, standardization, cen-tralization, and the qualities of commodities described in Chapter 2.

chapter two

The privileged qualities of commodities

Contents

2.1 The process of commoditization

The roots of commoditization lie in the effort to improve life through commerce and trade. The benefits of trade begin with the simple fact that raw materials, knowledge, and experience are unequally distributed across the Earth. Goods are stored, packaged, and transported so that what is relatively abundant and less desired here but relatively scarce and more desired there can be traded for what is abundant there but scarce here. Individuals and regions tend to specialize in productive activities for which they have special skills or access to resources. By concentrating on what one does best or most efficiently, growth leads to economies of scale, further specialization, and further unequal distribution of resources, which then encourages more trade. Systems of barter and trade emerge and are institutionalized, as a result potentially making everyone better off. Further institutional arrangements are then created to promote and facilitate exchange. Laws of commerce and contract develop. Corporations are created as legal entities with rights and responsibilities. Despite many diversions and contractions in the cycles of growth, the trend in the narrative history of trade and commerce is continually toward expansion. Once society grants the self-regulating market a determinative role in decisions regarding the allocation of resources, then the expansion of the economy operates in a classic feedback fashion so that over time its pace accelerates. In the end, both because the economy must generate wealth to attract interest-bearing investment and because human population grows, sustaining a high quality of life comes to require that the economy continue to grow. This growth means ever-increasing production and circulation of goods and services and therefore ever-increasing use and consumption of energy and materials and all the accompanying waste and by-product pollution.

As an economy develops people come to depend more and more on commercial sources for the necessities of life. In a modern economy people have more cash than they do skills of self-reliance and mutual aid. We enter the market with something to sell in order to obtain the resources we need to obtain the material conditions for our well-being. We can only be successful in that market if we have something to sell, even if it is only our time, energy, attention, and skills. As the market dominates the allocation of resources, people have little choice but to transform more and more of life into things that can be bought and sold. Repulsion at the crass commercialism and materialism of modern society may be a common theme of social criticism and a recurring story line in the popular culture, but the problem has little to do with the inherent acquisitiveness or greed of human beings and is instead part of the systems dynamics addressed by this book. Although greed, avarice, and dishonesty may all play their role in the growth of the economy, they are not necessary for commoditization to emerge as a determining factor in the structure of the economy. Throughout this book I attempt to show that the environmental crisis of our time is *not* anyone's fault but is instead a predictable outcome of the economic dynamics described here as commoditization. What *is* wrong, irrational, and avoidable is deliberately ignoring the unintended negative environmental and social consequences of runaway economic growth and commoditization. The fact that the consequences are unintentional does not make them inevitable. Economic reform that leads to a new economy in which quality of life steadily improves without continuous expansion of the physical dimensions of the economy may be possible. This can only happen by understanding the forces I describe as *commoditization* and developing policies that counteract them.

Commoditization is not an easy concept to describe. It is virtually agentless; there are no people whose job it is to commoditize, although there are scores of individuals whose jobs are to serve the commoditization process; advertisers in particular, designers, scientists, engineers, marketers, and investment specialists of all kinds. They (we) are operating, however, out of our best understanding of our self-interest, doing work that the economy rewards. In terms of the system dynamics of commoditization, there are no responsible parties. Commoditization operates gradually over time and may not be evident at any single moment in time or any particular place. It is a feature that both emerges out of economic life and then shapes it so that in the evolution of the economy its effects gradually intensify. Goods, services, and practices come into existence as the result of human ingenuity and the investment of time and resources to build a tool, shape a product, or invent a practice in order to solve a particular problem people face. Some of the resulting inventions and innovations are particularly marketable, others are not. If the system for rewarding investments of human time, attention, and resources are determined largely by the self-regulating market, then what will receive time, attention, and resources will be that which is likely to be able to compete in the marketplace, those goods and services capable of being handled and promoted as commodities.

A commodity is an object outside of oneself, a thing that by its qualities satisfies human wants of one sort or another. It is a useful thing, but mostly it is a thing, not a relationship. *All other, noncommercial possibilities for satisfying human wants and needs — common goods and services that involve interpersonal relationships or relationships between a particular place or ecosystem and the people who live on and in it — that can't easily be bought and sold will not receive anywhere near the same amount of time, attention, and resources invested. This simple fact has dramatic ecological and social consequences.* The qualities of a good or service that make it a commodity able to enter the marketplace (see Table 2.2) make it best suited for facilitating the accumulation of wealth and its abstraction in the form of money and other forms or representations of purchasing power, and at the same time, as we shall see, make it more destructive for the environment. And here lies the heart of our dilemma and the project of this book. Can we prosper while protecting the environment, or is commoditization such an essential part of wealth-production that any attempt to curb its excesses will inevitably impoverish us?

Commoditization is best understood as a process involving subtle economic pressures that "select" those options for satisfying wants and needs that are most suited or fit to enter the marketplace, meaning they mobilize optimal flows of material and energy for their own existence. In other words, they can successfully stimulate investment. Imagine an array of options to achieve any given purpose (any particular need or desire).

Commoditization in effect "evaluates" each option and picks those for fuller development that have the greatest potential to serve commercial ends. *Commodity potential* is a measure of the degree to which a good or service carries the qualities associated with a commodity. Goods and services can be described as having high commodity potential (HCP), or low commodity potential (LCP).

Everything has some commodity potential, even if, as in for example interpersonal relationships, it is very small. In an economy where resources are primarily allocated through the self-regulating market, all things evolve toward their maximum commodity potential, no matter how small or large that potential may be. Commodities are those goods (in the most generic sense) that can thrive under conditions that select for those qualities that distinguish a commodity from a noncommodity. If we want to understand how this works, we must first identify what those qualities are. Let us first compare the difference between goods and services that are highly suitable as commodities (those goods and services with highest commodity potential, HCP) with those that are less so (or LCP). This way we can begin to understand the qualities commoditization selects and why.

For example, pain medicine has inherently more commodity potential than does massage, insecticides more than pest management, increased energy production more than energy conservation, pollution control more than pollution prevention, Prozac more than counseling, computers more than teachers, cars more than mass transit, and on and on. Every good or service has some commodity potential, and in so far as it does, the selection pressures driving the evolution of a market economy will select for that potential, bring it out, and the good or service will evolve toward its maximum commodity intensity (Figure 2.1).

Hence it is not surprising to find that purveyors of books, seminars, and tools about organic farming fare better economically than do organic or traditional farmers themselves. This is very similar to the reason why the purveyors of pharmaceutical medicines and medical equipment fare better than the nurse's aide and other hands-on caregivers. They make considerably higher salaries. What we will discover in our exploration of commoditization is that over time as the process operates those people and communities most closely associated with LCP goods receive a small and decreasing proportion of economic rewards and constitute, across all sectors, an oppressed class of people. This will be discussed at greater length in Chapter 5.

There are two key differences between LCP and HCP goods and services, and both have to do with suitability for mass production and distribution. One has to do with the relationship between the good or service and a particular history, place, and culture. When something best provides a service in the context of a particular place or culture it cannot effectively be mass-produced. The second and related feature has to do with how much room

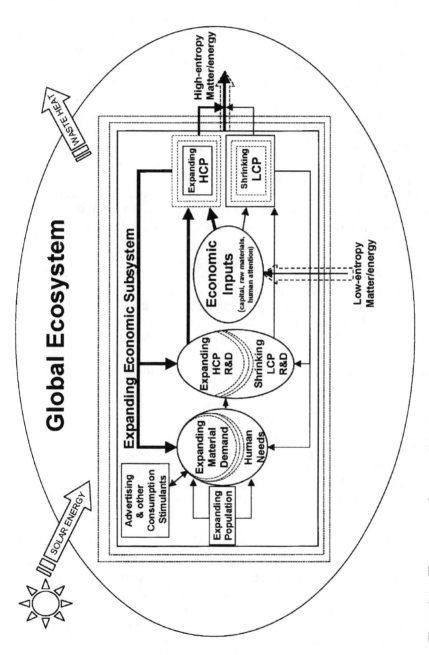

Figure 2.1 The process of commoditization.

there is for increasing the productivity of labor. For example, the productivity of direct services provided on a person-to-person or person-to-group basis has certain intrinsic limitations. A doctor can only see so many patients, a social worker so many clients. Commoditization still operates on these services, and they can and do become commoditized to the full extent possible. What we end up with when we settle for an economy that privileges HCP over LCP goods and services is increased labor productivity without the counterbalancing policies that protect the livelihoods of people displaced by such productivity.

The point is not that mass-produced commercial goods are bad and less-commercial goods are good, but that the propensity in the modern economy to gradually and inexorably marginalize and underdevelop noncommercial goods and favor commercial goods is a huge obstacle in the way of sustainability. Consider the amount of resources allocated to basic social services. Such services in general have low commodity potential. The United Nations Development Program estimates the added annual cost, over and above the current expenditures for basic social services in all so-called developing countries, to be as follows:

Basic education for all	$6 billion
Clean water and sanitation for all	$8
Reproductive health care for all women	$12
Basic health and nutrition	$13
Total	$39 billion

Compare these numbers with estimates of spending for selected highly commoditized goods:

Cosmetics in the U.S. alone	$8 billion
Ice cream in Europe	$8
Cigarettes in Europe	$50
Business entertainment in Japan	$35
Pet foods in Europe and U.S.	$17
Advertising in the U.S.	$101
Total	$219 billion[1]

Ideally, the economy would foster both sets of qualities: those associated with commodities and those with noncommodity values. At present, because commodities receive large investments of time and resources in research and development, they continuously improve, while noncommodities, which receive far less R&D, relatively stagnate. The solution to the environmental crisis is not economic or cultural stagnation or a return to some mythical conception of a "traditional lifestyle." The way out is clearly forward, through innovation and creativity, developing tools and techniques — both commercial and noncommercial — that meet human needs with the minimum amount of energy and materials.

2.2   Examples of how commoditization works

Consider, for example, children's need for play. At one end of the scale of commodity potential are such mass-marketed toys as Barbie dolls, super-hero action figures, and the packaged entertainment that accompany them. These products are inexpensive, marketed worldwide, involve immense sums invested in product research and development, packaging, and marketing.[2] From 1987 to 1997, worldwide Barbie sales more than tripled, from $435 million to $1.7 billion. Industry analysts estimate that on average, American girls receive three new Barbies each year and maintain a stash of eight or more.[3] Their production, largely in China, is energy intensive, fossil-fuel dependent, and involves the highly publicized exploitation of cheap labor in poor countries and mountains of industrial and post-consumer waste.

In the middle of the scale of commodity potential are those things that are usually locally produced and can be bought and sold but that involve some personal contact between buyer and seller: hand-crafted dolls, toys, and games, usually made from renewable materials and with local or cul-turally idiosyncratic designs. They are the goods of the crafts market and bazaar. Also in the mid-scale of commodity potential are all the services for sale: childcare, playgroups, clowns for hire, etc.

At the far end with the least commodity potential are activities that involve the voluntary allocation of adult time and attention and an interac-tive relationship between adults and children: making angels in the snow, play with found objects, group play, sing-alongs, and all the goods of inter-personal contact.

Ironically, despite the huge sales numbers, as in the case of children's play, noncommercial alternatives are often the more preferable from the individual perspective. Parents often prefer that their children play games with them rather than spend their days in video arcades; developmental psychologists agree that the commercial culture cannot provide the same child-rearing functions as human contact. Commercial commodities are often convenient substitutes for the nonmarket goods of adult time and attention. Sustainable development may have less to do with sacrifice than with *finding ways to get what we already want, which is each other.*

Through the commoditization of play, children are trained in the ways of consumerism. Imagination, the active engagement of the individual mind (a good with low commodity potential) is replaced with entertainment (a good with high commodity potential). Rather than active participants, chil-dren become passive consumers. They do not know where the material in their toys came from or where it will go when it is discarded. The person who made the toy has no relationship with the child. By contrast, when a child transforms found objects from the natural world into pretend objects of their imaginative world, what occurs is a creative relationship between a child's imagination and Nature. Increasing commoditization, as we will see in many of our examples, means a diminution in relationships between people and the Earth and people with each other. It also means the loss of the power of imagination. When the generic teddy bear is replaced by a character from a Disney movie, the child's imagination is replaced by a copyrighted character with a prepackaged history. When child's play is enhanced or replaced by commercial toys, it is not clear that the child's quality of life has improved. What is clear is that the child has now contributed substantially to economic growth, and herein lies the distinction between growth and development. Commoditization distorts development and transforms it into mere growth. We will see the effects of this in many different sectors of the economy.

Table 2.1 presents short examples from several economic sectors. These examples will be discussed in greater detail in Chapter 4.

The degree of commodity potential is important to keep in mind, more so than the absolute difference between commercial and noncommercial goods. It is neither realistic nor desirable to try to create a mass economy based exclusively or even predominantly on noncommerical goods and need-reduction strategies. What this book will argue is that noncommerical goods and services and need-reduction (and desire-reduction) strategies must have a place in a sustainable society and its economy. Without them, sustainability is impossible. Commoditization acts as a continuous force to privilege commodities over noncommodity alternatives and need-reduction strategies. What we need is a balance: a strong and healthy goods-producing market and a strong and healthy set of economic and social policies that counteract the effects of commoditization through conservation and need reduction. We should strive for an optimum in which the requirements necessary for human dignity and development are satisfied at a level of wealth that allows an advanced civilization to prosper without undermining the ecological basis of prosperity. This can be achieved by a society that does not allow commoditization pressures to distort social relations beyond the threshold of sustainability. If that means that in some or even many cases economic growth must be sacrificed for economic development, for increased community self-reliance and mutual aid, then social incentives and regulatory structures have to be put in place to achieve this. Under present circumstances in which a global self-regulating market system is emerging, the higher the degree of commodity potential the more likely is it to be developed to its full potential; the lower the commodity potential, the more likely it will remain underdeveloped, starved for society's attention, time, and resources.

The structure of an economy is largely self-organized through system dynamics, and its evolution occurs through operation of system-level forces. Adam Smith described the "invisible hand," the capacity for beneficial self-organization that emerges from the countless small decisions being made by individuals choosing to maximize self-interest. Like the achievements of similar self-organizing processes that lack an overall guiding intelligence, complex structure emerges despite the absence of intention. The process I call commoditization is one way to describe an aspect of the unintentional structuring forces that shape the economy (an invisible hand, if you will). To describe these forces

Table 2.1 High, Medium, and Low Commodity Potential Goods in Various Economic Sectors

Sector	High Commodity Potential (Commercial Goods)	Medium Commodity Potential (Artisan Goods)	Low Commodity Potential (Common goods)
Children's play	Barbie dolls, action figures, packaged entertainment	Handicrafts, childcare, live entertainment	Direct child-led interaction with natural surroundings, group play, interpersonal goods
Food production	Commercial fertilizers, pesticides, engineered seeds, mechanization tools, genetic material	Commercial manure, stored seeds, farm animals, tools for small farm, agricultural extension and research services.	Knowledge of soil, locally coevolved skills and techniques
Health care	Mass-marketed drugs, diagnostic equipment, hospital supplies, insurance	Practitioner-provided services, hands-on therapies and treatments	Knowledge of healing, personal health maintenance and illness prevention, life-style adaptations, sense of well-being
Energy	Grid-dispersed electricity, power plant equipment, fossil and nuclear fuels	Renewable energy sources, energy conservation services, wage labor	Personal energy conservation strategies, passive solar design, cooperative sharing activities
Transportation	Personal transport vehicles and the infrastructure of roads, etc., that supports it	Public transportation	Transportation reduction strategies (such as cluster housing near workplaces, etc.), walking
Environmental protection	Pollution control equipment, waste-to-energy incinerators and equipment	Recycling, pollution reduction/prevention services	Pollution prevention redesign, materials and energy use reduction strategies
Mental health	Mind-altering drugs	Professional counselors, coaches, fitness clubs	Peer counseling and mutual help, friendship, exercise
Finance/credit	Options, junk bonds, credit cards	Neighborhood banking, credit unions	Personal loans, gifts

Note: The process of commoditization favors those goods that have the quality of a commodity (see Table 2.2). Goods and services with low commodity potential (LCP) involve direct and cooperative relationships between human beings or between humans and the natural world. Goods and services with medium commodity potential (MCP) involve a direct exchange relationship between the purveyor of the goods and the end-user. High commodity potential (HCP) goods and services involve highly abstracted and usually distant relationships between producers and consumers.

as *invisible* and *unintentional* both clarifies and muddles our understanding of how economic forces operate. It clarifies by focusing us on the unintended consequences of the combined action of millions of individual decisions, but it confuses by distracting us away from the deliberate political decisions that determine the rules and the institutional structure of the economy within which these decisions are made. We make our purchasing and investment decisions based on our best understanding of our self-interest at the time of decision. We attempt to maximize our benefits and minimize our costs, even if we perceive those benefits and costs in noneconomic terms. The person who sacrifices personal economic gain for the good of his or her family, community, or the natural world is still acting in ways that maximize his or her individual benefits as he or she understands them.

If structural elements in the economy privilege certain kinds of goods and services with more investment of resources, time, and attention, then those goods and services will be more available and less expensive than their alternatives and consumers will choose the less expensive and more available goods and services. More sales then justify more investment, in a classic positive feedback loop. The resulting conditions appear to derive freely from free consumer choice, but it actually is a more complicated causation involving free consumer choice, yes, but consumer choice made in the context of conditions determined in part by initial investment decisions. These investment decisions are only partly determined by anticipation of consumer demand. There are many other considerations of potential profitability that only loosely relate to consumer demand. These considerations end up, as we will see, in privileging certain types of goods and services over others. If the net effect of that privileging is that our economy is unnecessarily destructive of the natural environment, no individual consumer has made a decision in favor of destruction. No individual producer has chosen to be deliberately destructive either; they have simply tried to minimize costs. Individual choice *does matter* and people can choose to make environmentally aware purchasing decisions and businesses can decide to act in an environmentally responsible manner, but the net effect of these individual decisions will likely be inadequate without changes in the system of preferences that favor the more environmentally harmful products. Thus the environmentally destructive aspects of economic behavior result in part from systematic distortions that encourage certain individual purchasing and investment choices in the direction of commoditization and discourage them from choosing alternatives. In the aggregate, this unintentionally shapes our economy and, as I will show, distorts economic development in ways that are enormously destructive to the ecological health and integrity of our world.

The goods and services so privileged are those that are most suited to being treated as commercial commodities because they are suited for buying and selling. Commoditization pressures act in the economy over time to gradually and inexorably expand the number of commercial commodities on the market, the geographic spread of their availability, and the range of types of needs and wants for which commoditized satisfactions are available.

In the process, worldwide they outcompete noncommercial solutions involving mutual aid, community self-reliance, and need-reduction strategies.

There is benefit to considering commoditization as a system feature. In this way we can begin to develop economic and social policies that counteract its effects. Consider for example the social choices made in the allocation of health care resources, including money, time, and attention. Some portion will be allocated to health promotion and disease prevention and some portion will be devoted to treatment and cure. It is apparent in both economic and health terms that disease prevention is preferable to treatment after the patient is sick. But prevention is a need-reduction strategy; it reduces the need for treatment. It runs counter to the pressures for economic growth, which depend on the creation and stimulation of need, actual and/or perceived. The negative effects of commoditization can be seen then in the health care economy where resources allocated to care and treatment far outweigh those spent on health maintenance and disease prevention (see Chapter 4).

Analyses of the energy economy point out similar distortions. There are obvious economic and environmental advantages to investing a dollar in energy conservation over the same dollar invested in producing new energy. But energy conservation is a strategy for <u>reducing need</u> and will succeed only if it can instead be in some way transformed into a market opportunity for the sale of energy conservation goods and services. Any and all true need-reduction strategies that benefit people by reducing their dependence on consumption are severely undervalued in a commoditized economy. There are similar examples in every sector where expenditures directed toward reducing or eliminating the need for a product would be far more efficient than the same expenditure in products designed to satisfy a need. The concept of commoditization explains why need satisfaction, in fact why the deliberate manufacture of desire, is invariably preferred in a growth economy over need reduction. By providing a cross-sectoral explanation for why treatment attracts more social and economic investment than prevention and energy production more than conservation and so on, the concept of commoditization helps create a conceptual basis for need-reduction policies that make social allocation of resources more rational and just across all sectors of the economy.

So what exactly are the qualities that distinguish goods with high commodity potential (commercial goods) from those with low commodity potential (public, common, and noncommerical goods)? What makes it so that one thing is easily bought and sold, and another is not? What is it about the goods that bear these qualities that helps them outcompete those that do not? How are these qualities related to the environmental impact of different goods and services? These are key questions which, as we will see, bear heavily on the prospects of developing alternative, earth-friendly economies.

Consider agriculture. The key factors involved in agricultural production are soil, sunlight, labor, seeds, equipment, farming skills, fertilizers, pest control products, land, fuel, water. In this list there is an array of highly commoditizable goods as well as common goods that are more difficult to commoditize and some that are in between. This book argues that an evolutionary propensity created by economic distortions called commoditization exists which, unless deliberately counterbalanced by public policy, leads toward ever-increasing emphasis on the most commoditizable aspects of farming and away from the least commoditizable. The impact on agriculture and the food industries of the privileging of commodities is discussed at length in Chapter 4.

2.3   *The attributes of commodities*

Table 2.2 compares the attributes of goods and services with high commodity potential with the attributes of those with low commodity potential. The third column suggests the social, economic, and environmental costs of privileging the qualities of commodities. The right-hand column suggests the social, economic, and environmental benefits of developing the qualities of commodities. The issue is not that noncommodities are necessarily preferable, but that a sustainable economy requires investment and social commitment in both commercial and public goods, both commercial solutions to human problems and noncommerical ones, both a strong private sector and a strong public sector, both the capacity to meet human needs with commercial goods and services and to meet them or avoid them with community self-reliance, mutual aid, and other need-reduction strategies.

In thinking about the characteristics of goods and services that become highly commoditized we need to think about the qualities that make a product attractive to private investors. We naturally look for a good chance for a healthy return on our investments. This basic requirement, to get back more than we put in, is a fundamental factor in the social and political commitment to continuous economic growth. Without it, profits are much less likely, and investment dries up. The fundamental ingredient of profitability is low costs and high return on sales, either through high prices or high volume. As the market economy expands until it incorporates the entire world, the meaning of high sales volume continues to grow along with sales potential. In order to compete, and provide an adequate return on investment, a good or service must have all or most of the characteristics described below. Brief examples are given of how, by systematically favoring these characteristics, commoditization distorts the economy and amplifies the strain it places on the environment.

2.3.1   *Ability to assign and protect property rights*

The ability to lay effective claim of ownership to something is the most fundamental determinant of its commodity potential. Effective ownership includes the ability to prevent others from benefiting from what you own and the ability to share or exchange what you own for something you want.

Table 2.2 Some Characteristics of Commodities the Economy "Selects" Over Those of Noncommodities, and Associated Effects

Attributes of Goods and Services with Low Commodity Potential	Attributes of Goods and Services with High Commodity Potential	Negative Effects of Commoditization	Positive Effects of Commoditization
Openly accessible: difficult to establish rights, widely available, difficult to accurately price.	**Appropriable,** excludable, enclosable, assignable: simpler to establish right of ownership, easier to establish price.	**Privatization** accelerates decline of sense of community and the common good and increases commercialization of all aspects of life.	**Releases** individual and corporate entrepreneurial energy.
Rooted in local ecosystem and community.	**Mobile,** transferable, easy to package and transport.	Propensity for mobility increases flows and export of energy and materials.	Makes trade possible, increases choice.
Particular, customized, decentralized and diverse: each culture potentially derives the best practices for its particular environmental context leading to diverse customized goods and practices.	**Universal,** standardized, centralized and uniform: adaptable to many contexts.	Reduces cultural and geographic diversity; standardized methods may not be suited to particular ecosystems, as a result efficiency potential is reduced.	Allows rationalization of production, economies of scale, and transfer of skills.
Systems-oriented: development occurs in the context of a whole system; goal is system optimization; product is developed to serve the system.	**Product-oriented:** development focuses on maximizing output of products; goal is profit maximization; system is developed to serve the product.	Discourages systems thinking, keeps attention on parts rather than wholes. Undermines capacity for ecosystem approaches to decision-making. Overdevelops competitive skills and underdevelops collaborative skills.	Produces cornucopia of products.

continued

Table 2.2 *(continued)* Some Characteristics of Commodities the Economy "Selects" Over Those of Noncommodities, and Associated Effects

Attributes of Goods and Services with Low Commodity Potential	Attributes of Goods and Services with High Commodity Intensity	Negative Effects of Commoditization	Positive Effects of Commoditization
Dispersed energy: energy is used and dissipated at the site of the activity or point of exchange or consumption.	**Embedded energy:** production is energy intensive. Packaging, transportation and promotion add to energy embedded in the product.	Concentration of energy causes ecological disruption at the point of its release. Commoditization of fuel facilitates dramatic increase in energy availability and use.	Increased energy production is tied to wealth production.
Dispersed knowledge and skills, convenience is not goal, use requires relevant knowledge and skills.	**Embedded knowledge or skills,** convenient. Use inherent in design and material.	Impoverishes knowledge base particularly at the personal, local, and regional levels.	Convenience frees human attention for other activities.
Low capital intensity High energy productivity High labor intensity Low labor productivity. **Consumption and use involve cooperative relationships**	**High capital intensity** Low energy productivity Low labor intensity High labor productivity **Consumption by individuals**	Eliminates jobs, encourages replacement of workers with fossil fuel energy. Promotion of individual consumption reduces the efficiency gains made possible by sharing, increases flow of energy and materials.	Increased productivity frees capital to invest in new productive activities creating new jobs. Increased individual autonomy and freedom.
More variable Unpredictable Unreliable	**More stable** Predictable Reliable	Predictability tends toward simplification, including loss of diversity and redundancy in ecosystems.	Increased predictability and reliability benefits all human activities.

Design follows and mimics natural flows and cycles.	**Design resists and/or alters natural flows and cycles.**	Failure to promote and use ecological designs leads to increased energy use and more waste.	Overcoming ecological contraints opens more possibility.
Concrete, tied to physical and biological constraints.	**Abstract,** less direct ties to physical base of reality.	Reduces knowledge and awareness of physical basis of human life and culture.	Overcoming physical constraints opens more possibility.
Coevolutionary and traditional, evolves in the context of specific ecosystem and culture, relationships often structured by custom.	**Path-breaking,** break from bonds of place and tradition. Relationships structured by contract.	Loss of traditions and traditional knowledge.	Overcoming cultural constraints opens more possibility.
Complex, involving multiple relationships, embedded in social and ecological web.	**Simple,** self-contained, extracted from relationships.	Simplification leads to general loss of diversity, redundancy, and resilience.	Narrowing focus to one or few variables makes production more manageable.
Long-term, stable returns.	**Short-term,** large return on investment.	Increases speculation, reduces investments in sustainable opportunities.	Increases wealth and its accompanying benefits.
Sufficient, optimal service for minimal expenditure of material and energy.	**Efficient,** the most exchange value for the least investment.	Reduces capacity to develop low impact living, accelerates commoditization.	Efficiency frees reserves for other wealth producing activities.
Contributes little to GNP, the less commoditized a good or service, the less it contributes to GNP.	**Contributes to GNP,** GNP growth measures commoditization.	Public policy goals become tied to growth in size of economy rather than improvements in quality of life.	GNP represents accurate measure of economic activity and is closely linked to improving quality of life.

A good can only be commercialized if a method can be found to exclude as much as possible from the benefits of a good or service those who do not pay for it. In addition, rights of property, if the property is to be commoditized, must be appropriable and transferable — property rights must be recognizable as attaching to the good or service and free to be assigned to someone or some authority, public or private, who is able to assert ownership rights and have those rights affirmed and backed by society. The easier this is to accomplish with any good or service, the more commoditizable it is. The stronger the claim to ownership the easier it is to exclude others, and vice versa. It is possible to own an air filtering machine or the manufacturing plant to produce such machines, and one can exclude from the benefits of air filtering machines all except those who pay the purchase price. It is much more difficult, on the other hand, to exclude from the benefits of fresh, clean outdoor air those who have not contributed to the cost of air pollution control. If I own the rights to produce an air filtering machine I can assign those rights to another in exchange for something in return. I cannot assign rights to clean air to any particular individual. The structure of an economy emerges as the result of innumerable economic choices. If the combined resources of an economy are preferentially invested in goods whose defining features are their ability to exclude all who do not pay, then the net result will be a society with far more boundaries than common space. It is no surprise then that among the great critiques about advanced industrial societies is the concern over the ways people are increasingly walled off from each other in private, guarded spaces, while less and less of life is lived in the common spaces where people readily interact.

Enclosure refers to the process of privatizing something that was previously accessed and used in common by a community or other recognizable group.[4] The paradigmatic example occurred when the value of wool exploded as the textile industry became a major economic force in late medieval England. The aristocracy "enclosed" or asserted exclusive ownership and use rights over vast tracts of grazing land that had previously been used in common by peasants for subsistence farming, grazing, and wild harvesting.

The story of economic development, economic growth, and commoditization is the story of the gradual expansion of property rights into more and more of life. Economic development is only likely to proceed where it is clear who will benefit from the development. Development requires investment with all its accompanying risk. Risk is greatly increased where the property rights to the thing being developed are unclear or unprotected by the institutions of society, such as the police and courts. This is why effective and enforceable commercial law is considered a prerequisite for economic development.

In theory anything can be commoditized. Some things, like genetic information, Web sites, or a slot in geosynchronous orbit in outer space, come into existence as commodities as a result of scientific, technological, and institutional achievements. Ideas and innovations become property through advances in patent, copyright, and intellectual property laws. No individual or corporation may claim as one's own or has the sole right to commercialize much of what we value most in life — friendship, good air, clean water, a sense of well-being, cooperation, civility. However, this class of goods that lies outside the potential reach of privatization has continued to shrink over time. There is a constant pressure to commoditize whatever can be commoditized and to create distorted facsimiles of that which cannot be. Things like friendship and feelings of well-being can only be incompletely commoditized through personal call-in lines, paid counseling, or mood-altering drugs. Economic "progress" is largely the tale of clever innovations that attach property rights to things that previously could not be owned or sold.

Looking at our children's play example, it is relatively easy, given an effective international intellectual property rights regime, to assign ownership rights not only to the inventory of Barbie dolls themselves but also to the design, the concept, the name, the accessories, and so on. The capacity to abstract property rights and treat them independently to some extent from the thing itself is associated with a good or service with high commodity potential (HCP). With simple toys made of readily available material, ownership is likely to extend only as far as the toy itself not to the design or the concept. Thus, what characterizes products with mid-range commodity potential (MCP) is the fact that ownership extends to the good or service itself at a particular place and time. There is necessarily a physical relationship between the owner and the property. Finally, the various forms of children's play, such as making angels in the snow, snuggling, etc., are difficult to commoditize and are examples of goods and services with low commodity potential (LCP). The only way they can be commoditized is through the sale of the service provider's actual presence and attention.

Commoditization operates across the full range of economic sectors. Even in LCP sectors such as children's play, books are written and copyrighted on how to play with children, patterns are produced and distributed for making homemade toys. As we will see, no matter how you subdivide the economy the most commoditizable sector always receives more investment and is more highly developed than the least commoditizable, no matter how small a portion of the overall sector the commoditizable portion represents.

Those aspects of life that lend themselves to establishment of clear property rights benefit from economic investments and those for which property rights are difficult to assign remain underdeveloped. The direction of technological development as influenced by commoditization steadily moves things into the category of property. Technology is aided in this by the laws and institutions of patents and patent protection.

Patents in agriculture, up until the creation of hybrid seeds, were limited to innovations in farm equipment and the chemical formulae for commercial pesticides and to a lesser extent fertilizers. The thousand years of development through which agricultural crops were steadily improved did not and could not involve patent protection, because anyone possessing the plant or seeds had the means to reproduce the product, and thus the great indigenous agronomists of the centuries could not appropriate to themselves the monetary value of their successes. As a result, the commoditization of agriculture remained incomplete. With the development of hybrid seeds that reproduce poorly, if at all, it became possible to create plant varieties with commercial value and to provide patent protection to plant breeders under the Plant Patent Act. As a result, most private-sector breeding efforts went into hybrids. Until 1980 it was not possible to patent newly created biological organisms, but then the U.S. Supreme Court authorized patents for genetically engineered microorganisms. This was gradually extended by 1987 to plants and animals as well. One result of these developments is that privately funded research grew roughly 20-fold between 1960 and 1992, much of this going into hybrid research and then biotechnology.[5] By 1994 intellectual property rights had been extended to more than 3000 new crop varieties, with private industry holding about 87% of them.[6] Historically, plant breeding research aimed at improving yields was funded mostly by government agricultural agencies. In recent years private R&D has outpaced public.

Social knowledge and skills that grow out of the relationships between many people or between people and the natural world (such as how to organize communities, how to maintain the soil in a particular location) are very difficult to assign as property. These kinds of goods and services, with inherently low commodity potential, will be developed only by society acting collectively, through its social institutions. In practice, this type of development has occurred throughout the world and throughout history through common property arrangements or agreements between groups of owners of communally owned property.[7] But most community-based common property arrangements are less effective at commoditization because their use and production is tied to a specific community.

Without a deliberate approach to public investment in noncommoditized goods, these areas of the economy — and they exist in every sector — will remain considerably underdeveloped in comparison with their private-property counterparts. In general, in the modern industrial state, commoditization distorts public investment toward investment in and purchase of the most commoditized approaches to public welfare. Public expenditures for housing, health care, prisons, education, military, and public welfare are increasingly spent as payments to private interests and industries.

When goods are produced by the cooperative effort of many different people it can be difficult to determine ownership rights. Large-scale industrial production could have been hampered by this potential confusion about ownership. It was the effort to resolve this problem that led to the modern invention of the corporation, which can, like individuals and families, lay claim to rights of ownership and which can petition the legal system to uphold its rights and remedy its grievances. As we will see later in Chapter 8, economic policies that counteract the negative effects of commoditization will, among other things, expand the economic role of nonprofit corporations, cooperatives, and other forms of collective organization that can promote noncommercial ends.

In the process of commoditization, increasingly, things that were once held in common or were provided as a result of interpersonal relationships or were avoided by prevention get transformed into private goods with property rights assigned to individual or corporate owners. One of the results of this ongoing commoditization is that much entrepreneurial energy and invention goes into transforming the commons into private property.

Often the less valuable (or least commoditizable) aspect of public property is detached from the most valuable (commoditizable) which is then assigned to private parties. This is what happens in the sale of mineral and logging rights on public lands. This has led to some of the most outrageous filching of the common heritage in public land.[8]

2.3.2   Degree of mobility and transportability

The second most important feature that distinguishes something with high commodity potential is mobility. Property rights may be a prerequisite for commoditization, but without transportability the ability to exchange is limited by geography. There are two main types of transportability: the ability to transport the good itself with ease over large distances, and the ability to transfer and transport property rights when the property itself is less physically transportable. The first is associated with the capital accumulation of the merchant, the latter with the capital accumulation of the financier.

A great deal of economic activity is devoted to packaging and transportation. This is a major component of energy and materials use in the modern economy. Consider again our previous examples of the air filtering machine and the Barbie doll and consider how transportability is an important aspect of their commodity potential. You can package and transport the air filtering machine; you would have a much more difficult time doing that with the air. Barbie dolls can travel anywhere; playtime happens only at a particular time and place. Some things such as real estate cannot be transported; instead the increased mobility of people expands the number of times homes change hands. The more mobile products and customers are, the greater the potential market opportunities. Mobility of both goods and customers greatly expands potential markets. As a result, the tools and techniques of storage, transport, preservation, and advertising have become highly developed in a modern economy. Recently the development of Internet commerce has greatly increased the mobility of customers as well as made information about larger numbers of products vastly more accessible.[9]

The increased mobility of financial capital has also greatly transformed the world economy, as it grows ever easier and faster to move money in various forms instantaneously to virtually anyplace on Earth. In the evolution of the economy, commoditization creates a selection bias for mobility. For example, investments in agriculture and food processing often focus on improving the technology of storage, packaging, and preservation rather than nutritional content or soil health.

As a result of this bias for transportability and mobility, that which is site-specific or simply has deep roots and ties to local communities loses out to more readily transported and transplanted goods in the competition for investment. The skills and attitude of stewardship associated with long-term knowledge and care of a particular place or community can be considered economic goods and services in themselves. They are exactly the sorts of goods and services against which the bias for mobility acts.

The growth in package delivery not only increases commodity consumption but in general breaks down the personal relationship between buyer and seller. Such relationships, because they limit mobility, become obsolete in a highly commoditized growth economy.

2.3.3   Degree of universality and particularity

Closely related to the quality of transportability is universality or standardization, the ability of a good or service to serve nearly the same function wherever it goes. "Transportability" means the capacity to separate a good or service from its particular physical relationships with its surroundings and move it elsewhere while still maintaining the product's physical integrity and usefulness. The effects of the privileging of transportability are critical to our understanding of the ecological consequences of commoditization. What is decidedly not transportable and therefore underdeveloped as a result of commoditization is anything that has its value rooted in or dependent on a particular context, anything that is directly tied to a particular place or culture. The privileging of transportability favors those things that can serve equally well in any culture or climate. Thus we have the privileging of those things that lend themselves to standardization.

The link between standardization and transportability is beginning to break down as a result of advances in communications and computer-aided design. It is becoming increasingly possible to fine-tune productive processes in real time and produce a multitude of culturally and individually specific variations on standard products. This, in effect, increases the trend toward the commoditization of cultural and individual difference where consumer preferences become the defining feature of culture and community.

Consider what it means to privilege those types of goods and services that can readily be standardized. As soon as someone begins to include ecological considerations in product and service design criteria, one faces the inherent conflict with the economic bias for standardization. The building that is designed to take best advantage of sunlight, breezes, and trees is designed for a place and a set of conditions. Energy conservation efforts are unique to each set of circumstances. Organic farming is based on the unique conditions of soil, hydrology, native plant communities, and other site-specific considerations. Modern industrialized, commoditized agriculture demands standardization of farm methods and resulting crops. Since all ecosystems are inherently and specifically local, to design a tool or service to be ecologically fit is also to make it in some ways less transportable and universal. This conflict between the necessarily local and contextual focus of ecological design and the characteristically universal and standardized nature of most successful commodities has enormous environmental consequences. If we wish to design for our proposed definition of efficiency, meaning deriving the most service from the least input of energy and materials, then design must take account of the environmental conditions in which the product will be functioning. Otherwise, it costs material and energy to adjust to and operate under a wide range of conditions.

One example of this effect can be observed in the decline of cultural and agricultural diversity, and particularly in the decrease in crop varieties that make up the majority of the diet worldwide. At present, 90% of the world's human food supply comes from only 15 plant species and 8 animal species among the 10 million species of plants and animals in the world. The percentage of the total that is derived from the several thousand other edibles continues to decline.[10]

Many ancient cultures developed a wide variety of crops, each meant to produce optimal yield under specific conditions of climate and soil. Starting thousands of years before the Incas, the natives learned how to produce extremely high yields of potatoes from small plots of land. In the modern world, producing high yields has come about primarily through developing plants that can grow in different types of environments and, when necessary, through the manipulation of the immediate environment of the plant to ensure that it has just the right amount of moisture, nitrogen, and other requirements for maximum growth. Peruvians seem to have approached the problem in the opposite way. They sought to develop a different kind of plant for every type of soil, sun, and moisture condition. They prized diversity. They wanted potatoes in a variety of sizes, textures, and colors, from whites and yellows through purples, reds, oranges, and browns. Some tasted sweet and others too bitter for humans to eat, but the latter were useful as animal fodder.

They did not seek this diversity merely for the aesthetic pleasure of having so many shapes, colors, and textures, but rather for the practical reason that such variations in appearance also meant variation in other, less noticeable, properties. Some potatoes matured fast and some slowly, an important consideration in a country where the growing season varies with the altitude. Some potatoes required a lot of water and some required very little, which made one variety or another more adaptable to the highly variable rainfalls of different valleys. Some potatoes stored easily for long periods of time, others made excellent food for livestock. (Similar variations were cultivated for corn, quinoa, amaranth, and a variety of tubers with no English names.) At the time of the Spanish conquest, Andean farmers already were producing about 3000 different types of potatoes.[11]

Absent European conquest, indigenous American farming could well have continued to develop along these diverse lines. With the evolution of the modern economy in part through commoditization, the development of agriculture is no longer driven by local considerations of maximizing nutritional yield under conditions of ecosystem constraints but by the demands to produce agricultural commodities for a global mass market. As a result, the horticultural skills that have evolved over thousands of years in indigenous cultures everywhere become systematically underdeveloped, starved for the time, attention, and resources needed for development because locally specific products cannot be standardized and marketed successfully to a mass global market. It is possible, as it has come to pass in modern-day Peru,[12] that some indigenous crop varieties survive because they can claim higher prices in urban areas as specialty foods, but these prices keep traditional varieties of potatoes out of the economic reach of the native descendants of the people who's horticultural genius first developed them. As we will see, goods and services with low commodity potential often transform through commoditization into high-priced specialty goods available only to the wealthy. In fact, much of the modern commercial economy can be placed into two sectors: the mass produced, mass-marketed goods and services with high commodity potential and the high-priced specialty goods market, including high-quality crafts, handmade goods with unique designs, and personal services involving skilled labor.

Product standardization implies the capacity to expand markets and significantly reduce product diversity. The end result in any single location might appear to be greater diversity, as products from all over the globe become available to anyone with purchasing power. But in the aggregate there is a significant decline in diversity. Rather than millions of local and regional markets each with its own range of thousands of unique regionally produced goods, commoditization drives economic development to a single or few huge markets with far more products than any regional market but far fewer than the sum of all regional markets. The decline in market diversity creates greatly increased power in those global centers where economic decisions are made. Global economic power becomes increasingly centralized. This power is then used to support public policies that accelerate the commoditization process. This positive feedback mechanism is one of the ways in which the effects of commoditization increase at an accelerating pace over time.

Franchising is one result of the commoditization of mid-range commodities. Recall that goods and services with mid-range commodity potential are those that require some relationship between buyer and seller. This is true for much of the retail industry and such things as food service and restaurants where the nature of the product requires freshness and personal service. A franchise is a way of standardizing products and services in these mid-range commodities. The increase in the number of franchises is a surrogate measure for the standardization aspect of commoditization. According to *Entrepreneur Magazine*, in 1980 there were 134 Subway franchises; in 1999 there were 13,400. In the same period McDonalds grew from 5,749 restaurants to 16,000. The greatest growth in franchising came in the service industry. Franchised photography shops grew from 9 in 1980 to 965 in 1999. Training center franchises grew from 267 to more than 21,000. In 1999 there were more than 400,000 franchised outlets accounting for nearly a third of total retail sales in the U.S.[13]

2.3.4 Systems vs. products

There are many ways to address a problem. Commoditization has the effect of focusing attention disproportionately on products rather than systems. Without the pressures of commoditization, solutions to the problem of how to meet the needs of individuals and society would be designed to satisfy or eliminate the need with the least amount of time, materials and energy cost, and social and environmental disruption. Such solutions would necessarily focus on systems of problem solving and service delivery: health care systems, transportation systems, food production and distributions systems, housing systems, etc. These systems would consist of products, but also need-reduction strategies, energy and material efficiency, and systems of mutual aid and cooperation. Commoditization results in society consistently privileging product development over the other components of a systematic approach to problem solving.

As we have seen, the commodity potential of a product depends on it being identifiable in the singular as property, mobile enough to be exchanged, and standardized enough to serve its purpose under many different conditions. In other words, commodity potential depends to a great extent on detaching things from their surrounding ecological and cultural contexts and placing the focus of their development on the individual product rather than on the systems of which the product is one component part. As a result, development is focused to a far greater extent on health care products than on health, on environmental products rather than ecosystems, on vehicles far more than transportation systems, on farm products far more than farming. Commoditization keeps the focus of development on the production and consumption of individual goods rather than on the developing systems of service.

Consider the following examples of how development might differ with a systems focus rather than a product focus. A systems approach to agricultural development would have as its focus the relationship between ecosystems and crops. The health and long-term productivity of the soil would be at least as important as, if not more important than, the crop itself. Many different varieties of crop plants would be developed, each uniquely suited for a particular set of ecological conditions: drylands, mountains, small plots, and urban gardens. In other words, while the crop varieties would be potentially less damaging to the environment and require less inputs of commercial fertilizers and pesticides, the market for each variety would be limited to specific environments. If, however, development is focused instead on the production of super crops that can produce standardized high yields under many different conditions, given the addition of commercial fertilizers and pesticides and other inputs, then not only can the seed be widely distributed but products for artificially enhancing growing conditions also find a huge market. With the one approach, varieties are manipulated to suit ecological conditions, with the other the ecological conditions are manipulated to suit the crop. The focus of the systems approach is on the soil, weather, plant system, while with the commoditized approach the focus is on the final product. The knowledge of genetic manipulation that yields the products of biotechnology has far greater commodity potential than does the knowledge required for breeding plants that thrive in unique harsh conditions. Ideally perhaps, the focus would be on both, but under the economic conditions distorted by commoditization, R&D is so disproportionately concentrated on products that agroecological systems science is dwarfed in comparison and, in effect, deliberately underdeveloped.

Consider the development of the building construction industry. In order to design a building to be as energy-efficient as possible, architects and designers must think not only of pipes and fans but also about the construction materials and the lay-out of the building in relation to its surrounding environment; its orientation to the sun and prevailing wind patterns. If energy efficiency is to be designed into the building from the beginning then everyone must work closely as a construction team rather than as separate contractors for plumbing, lighting, heating, and so on.[14] This requires thinking about the building as a whole system involving energy flows in multiple directions. This systems knowledge is not readily transportable to other buildings. The specificity of the skills and practice limits their commodity potential.

Systems orientation involves a wider set of variables than does product thinking.[15] It is more complex. It involves multidisciplinary collaboration between people with diverse knowledge and skills. It takes more time. It also involves a sense of cooperation with Nature and requires deep knowledge from both empirical and experimental observation of the natural world.

The virtues and benefits of systems thinking have been extolled by many authors in diverse fields. There are the proponents of holistic health care, ecological engineering, organic agriculture, family systems therapy, and on and on.

The economic dimensions, particularly the bias in R&D expenditures toward products over systems, is almost always ignored in these analyses. It is as if all that was required was for scientists and scholars to begin thinking in systems terms. But science is as distorted by commoditization as all other social phenomena. The science rewarded with the most research and development support is always that which is most transferable to commodity technology. This is discussed in more detail in the next chapter.

Under the pressures of commoditization the only systems thinking that predictably gets supported are the technical and social systems that exist to serve the commoditization process. Systems thinking flourishes in the design and development of computers, communications networks, product supply and distribution lines, and the integration of manufacturing and sales of any particular good. The skills involved in large social and ecological systems analysis are typically compensated at a lower rate than the systems skills involved in product engineering and marketing. It is not surprising that the urban planner and the watershed manager receive far less financial compensation than does the systems designer in the computer or auto industry.

2.3.5 Cooperation vs. competition

Like systems thinking, collaboration (or consciously behaving as part of a system designed for achieving shared ends) is encouraged mostly within the context of the production of those products with the most commodity potential. Teams of people are brought together and handsomely rewarded to produce the latest fashions in automobiles, clothes, designer drugs, and movies. In our economy it is much more difficult to bring teams of people together to achieve noncommodity goals. When these teams do come together they find themselves with few resources and little power. Collaboration requires leadership services and organizational costs. In a commoditized growth economy these costs must be justified by the bottom line. Solving social and environmental problems and meeting social and environmental needs that don't lend themselves to product-based solutions requires systems collaboration. These collaborations require as much or more support than do product development teams, but their access to resources are limited unless they commoditize their approach and involve private partners who stand to gain. This is what happens in the defense and other public institutions, which purchase huge numbers of products from private parties using public resources. To the extent that public goods can be commoditized, they are able to increase their access to resources in a commoditized growth economy.

Most of the collaboration that does occur in a commoditized growth economy and social system takes place within a context of competition between alternative products. In this competition, noncommodity alternatives have little chance of success. The competition occurs between product teams. Winning the competition is handsomely rewarded. As a result we have created a bizarre set of social expectations that simultaneously extolls the virtues of competition *and* teamwork. Teamwork is systematically encouraged when it involves competing against other teams, communities, corporations, or nations.

Often the process of commoditization is associated with moving from collaborative types of institutions to competitive ones. There are many myths associated with the benefits of competition. Think about how often in recent years government services, schools, hospitals, and nonprofit organizations are encouraged to become more competitive. The evidence, however, is overwhelming that cooperative efforts, particularly when coordinated toward strategic and systemic ends, are more efficient and effective. In his book *No Contest* author Alfie Kohn argues that the belief that competition improves performance is ideological rather than based on the available social science evidence. In general this ideology has been used to justify the increasing privatization and commoditization of formerly public institutions in the name of competitiveness. Kohn quotes social psychologist Morton Deutsch summarizing the social science research which concludes that:

> Cooperative, as compared to competitive, systems of
> distributing rewards — when they differ — have more

favorable effects on individual and group productivity,
individual learning, social relations, self-esteem, task
attitudes and a sense of responsibility to other group
members. This conclusion is consistent with the research results obtained by many other investigators in
hundreds of studies. It is by now a well-established
finding, even though it is counter to widely held ideologies about the relative benefits of competition.[16]

"What does lead to excellence then?" Kohn asks. "This depends on what
field and task we are talking about, but generally we find that people do
terrific work when (1) they are inspired, challenged and excited by what
they are doing, and (2) they receive social support and are able to exchange
ideas and collaborate effectively with others."

Other than the reference to improved productivity, the other "goods"
promoted by collaboration — individual learning, social relations, self-
esteem, task attitudes, and a sense of responsibility to other group members
— all serve nonexchangeable, noncommodity goals, and in the competition
for investments of time, attention, and money they tend to be outcompeted
by commoditized goals. There is one exception to this rule, and that is in
the teamwork considered beneficial for product and other forms of corporate
development. This is why team-building consultants and personal coaches
of the world receive a welcome hearing in corporate boardrooms and training centers even while the traditionally cooperative public institutions are
being urged to become more competitive. Commoditization creates the conditions in which collaboration and systems thinking is encouraged only
when these skills are focused on the production of commodities. When these
same skills are focused instead on public and community goods and systems
that lie outside the effective control of private interests, such as management
of particular ecosystems or saving endangered species, then they are made
increasingly marginal by the force of commoditization.

2.3.6   Degree of energy concentration

In general most increases in productivity and wealth in developed economies have been the result of the exploitation of increasing amounts of
nonrenewable fossil fuels,[17] largely as a result of replacing human and
animal power with industrialized sources of energy (fossil fuels, hydro-
power and nuclear power).[18] We will discuss energy and commoditization
in greater detail in Chapter 4. For now we can note that commodity potential is closely associated with what is sometimes called "embodied" or
"embedded" energy. Goods and service skills with low commodity potential rely on far less direct energy input. Note the qualities that define
commodity potential — mobility, capital intensity, standardization,
packaging, and the use of synthetic materials — all require large inputs
of energy, usually in the most commoditized form of fossil fuels. These

fuels represent the work of thousands of years of plants' concentrating the sun's energy into a compact and volatile form. Fossil fuels are the commodities that make commoditization possible. Consider the example of agriculture. The industrialization of farming has meant large increases in the amount of food produced per farmer and farm worker; in other words, the labor productivity of agriculture has increased steadily. This has resulted in replacing the stewardship skills used to manage the small farm with huge amounts of energy in the form of fertilizers, pesticides, mechanized irrigation, and heavy machinery, all of which are derived from or depend on fossil fuel. In the U.S. the agricultural economy uses approximately 10 kcal of fossil fuel for each kilocalorie of food consumed, or about a gallon of petroleum to feed each person per day.[19]

While the intrinsic quality of agricultural soils has been in decline worldwide, yields continue to rise with increasing inputs of energy. With loss of soil quality and farming skills, the world has become increasingly dependent on high-energy inputs. This scenario is played out in other sectors as well, where commoditization destroys the skills and resource base for local self-reliance, and individuals and communities grow ever more dependent on imports of concentrated energy in the form of purchased commodities. This is also true for our other examples. Each Barbie doll, all the other plastic dolls, and all the packaging represent a concentration of energy in the form of plastic derived from fossil fuels. In many ways this concentration of energy in a mobile package is a characteristic of goods with high commodity potential. The increased mobility and packaging associated with commoditization adds additional steps between producer and consumer, and at each step material and energy is lost (converted to waste). Because of this, efforts to reduce energy use that do not involve a strategy to introduce and maintain countervailing forces to the pressures of commoditization are destined to fail. The policies needed to institutionalize these countervailing forces will be discussed in detail in Chapter 8.

2.3.7   Embodied knowledge vs. user knowledge

Closely related to mobility is convenience. Think about what convenience means. A convenient good or service is ready to use with minimal contribution and participation on the part of the user or consumer in terms of effort, knowledge, or experience. A good or service is more highly commoditizable if the knowledge that has gone into its development adheres in some way to the product. All the knowledge the user needs to produce music is embedded in the CD player and disk when it is purchased. The user brings nothing to it in the way of skill. CD players and CDs, by this criteria, have greater commodity potential than do guitars, which require considerable value added in terms of knowledge for use. All the knowledge associated with the chemistry of insecticides is embodied in the product, whereas alternative pest management strategies require observation and skills on the part of the farmer/gardener, and therefore chemical pesticides have much greater commodity potential than does Integrated Pest Management. The more the user must bring to the product in terms of attention and skill, the lower the commodity potential of the product.

2.3.8   The substitution of labor by technology

A person's labor has considerably less commodity potential than do machines and other forms of capital. It is difficult in modern times to press claims of ownership to another human being, so property rights cannot be asserted or transferred. Labor, although dramatically more mobile than in the past, is still less so than equipment and resources. Goods with low commodity potential generally have low capital intensity and high labor intensity, whereas the opposite is true for goods with high commodity potential, which have high capital intensity and low labor intensity. Since under the pressures of commoditization the trends in the economy move from low to high commodity potential, over time the investments in the production of goods and services steadily shift away from labor and toward capital. Commoditization creates a class feedback mechanism that gradually increases power of capital relative to labor, which increases the power of the groups that stand to gain the most from increasing commoditization. Figure 2.2 shows the trends in amount of capital vs. number of work hours per unit of dollar value added in the U.S.

2.3.9   Community cooperation vs. individual consumption

Commoditization favors the satisfaction of human needs through individual consumption of products. Alternatives to commoditization emphasize the satisfaction of human needs and the solving of problems through cooperative relationships among people and between humans and nature. Goods and services with low and mid-range commodity potential involve some significant and necessary relationship between user and place or between provider/seller and user/consumer. Neighborhood craft markets or bazaars are the marketplace for goods with mid-range commodity potential. Low commodity potential goods and services often involve cooperative efforts to build and maintain collective facilities, raise barns, and share information.

Increasing commoditization tends to break down the relationships between buyer and seller along with all other relationships. Less commoditized alternatives are often communal and cooperative, involving highly ritualized relationships with specific resources and a complex arrangement for sharing work and produce.

2.3.10 Stability and predictability vs. fluctuations and surprise

One of the most important goals of economic development is to bring stability and predictability, and with them increased personal and community protection against the vicissitudes of the natural world and economic life.

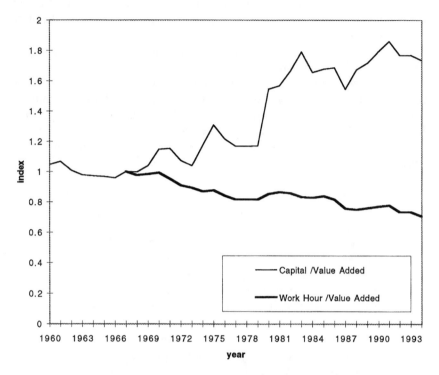

Figure 2.2   Index (1967 = 1) of capital, and work hour per value added by manu-facturers. (*Source:* United States Statistical Abstracts.)

This is particularly true of the advances in infrastructure upon which development depends. Much of the benefits of development — control, continuous yield, highways that do not turn to mud in the frequent rains, activities that happen and organizations that function because people synchronize their activities and reliably show up someplace all at around the same time — derive from these efforts. But like all benefits, some costs are involved. The attempt to control and limit as much of the variability of life as possible leads to homogeneity and the loss of diversity. The diversity of life is a result of different biological strategies that have evolved often to take advantage of different stages of regularly fluctuating cycles. Water level fluctuations are a good example. When rivers are forcibly contained in channels, flood plains stay permanently dry. As a result the vegetation that relies on occasional floods for reproductive or growth activities die away and along with it the floodplain ecosystem. Diversity and homogeneity are opposite and conflicting values. Like all conflicting values both cannot be maximized at the same time. But each can be optimized in the environment that includes its opposite. This can only be done when natural cycles are understood and development occurs with allowances for plenty of natural variability. This requires long-term observation and appreciation of particular ecosystems and their processes, a willingness to forego efforts to completely control Nature, and a capacity to restrict development. Each of these are undermined by the forces of commoditization.

Societies and economies exist almost by definition to reduce the impacts of unpredictability and extreme variability as much as possible. It is what lies at the heart of the matter of the human impact on the natural environment. All energy development is about evening out as well as increasing the availability of power, making energy usable. All flood control is about managing unpredictability and variability in water flows. All agriculture is about creating niches for favored crops and destroying niches for competing, less useful weeds. As Richard Norgaard explains:

> Modern agriculture almost always involves reducing the number of species in the ecosystem, typically lowering the combined efficiency of nutrient cycling, higher but less stable rates of production, and low biomass stocks relative to the ecosystem before modern agriculture. As people push ecosystems to meet their own needs they often intervene in some of the nutrient cycles and disturb equilibrating mechanisms which had previously evolved in the system. Fertilizing and management of legumes to replace lost nutrient cycles, weed control to offset successional processes, and insect control to compensate for displaced equilibrating mechanisms such as pest predators will probably have to be introduced. These new activities within the social system are costs.[20]

Fundamentally, various stressors increase diversity and complexity. Human efforts to reduce stress also simplify systems and make them more rigid. Human rational desire to control extremes of variability and randomness also inevitably lead to depletion of biodiversity and loss of ecological complexity (and with it, resilience and flexibility) overall. Therefore, the effort to preserve ecosystems is about modeling human development as best as possible on the natural cycles and variability in nature. This requires, in essence, designs for social systems unique to each ecosystem, economic activity best suited to utilize the available flow of nutrients and energy with the least disruption to their patterns. But this comes right up against commoditization. Ecologically appropriate development would require solutions to human problems (products, services, techniques) that are relatively specific to ecological conditions. Such specificity decreases the capacity for standardization and increases the need for skilled labor which decreases the commodity potential of the associated goods and services. But commoditization means that such goods are systematically selected out from the economic landscape. The problem with commoditization and its influence over the allocation of human attention and resources is its wild success at imposing human-valued order onto materials and energy and thus causing the gradual simplification of the ecological systems involved.

The human economy grows more simple over time despite the appearance of the opposite, that life grows ever more complex. The key to this seeming contradiction lies in the proliferation of commodities. The total number of commodities available to individual human beings increases as commoditization proceeds; hence the seeming multitude of information, consumer goods, entertainment, interrelationships with people all over the globe, etc. But from the perspective of the biosphere the economy grows increasingly simple as standardized commodities reduce cultural and ecological specificity.

Culturally specific common goods are part of complex interrelationships involving the coevolution of artifacts and techniques with ecological conditions. Consumer goods are much simpler. They replace embedded knowledge and history with embedded energy. Commoditization is a universalizing force. It gradually causes the replacement of a pluralism of locally appropriate goods and services with goods that have the quality of commodities. This is at the heart of the significant loss of cultural diversity as the planet develops toward a single mass culture served by a mass entertainment industry. The great diversity of locally appropriate ways of life, artifacts, and tools is reduced.

2.3.11 Design with vs. against nature

There are many ways to take account of Nature in the design and construction of products. There are architectural designs that accommodate themselves to the landscape and partly utilize solar energy and trees for heating and cooling. There are ways to produce materials that will eventually break down into harmless compost after use. There are industrial processes that model themselves on food chains where the waste from one process becomes food or input for the next process, eliminating waste throughout the cycle. What all of these have in common is the process of designing within a natural context. Context and relationship, as we have seen, limits the commodity potential of any good and service by increasing the need for skilled labor and attention to systems dynamics while limiting the capacity for standardization and intellectual property claims.

2.3.12 Concrete vs. abstract

One of the results of commoditization is a growing level of abstraction. As I have argued above, products, services, and accompanying lifestyles that are divorced from the particulars of relationships tend to be simpler to commoditize. This is one of the major reasons why we have such highly developed industrial technology and such poorly developed relationship skills with each other, with our communities, and the earth. The gradual commoditization of communication and entertainment encourages individuals to spend far more time watching television or surfing the Net than they do with each other. Time with each other is very difficult to commoditize, time with machines is easier. The ardent development of Virtual Reality points out a characteristic of commoditization: Virtual Reality is commoditizable, actual reality is not. In general, products and services that are abstracted from the physical bases of living reality are easier to commoditize than anything that is tied to physical and biological reality and constraints. As a result, commoditization tends toward a general reduction in awareness and intimate knowledge of the physical and natural world, with all the accompanying dangerous loss of ecological awareness and wisdom.

2.3.13 *Tradition vs. innovation*

It's a common misunderstanding to think that all innovation is a destroyer of tradition. In fact, all cultures continually face new challenges and encounter new information and develop unique responses to each new input based on their own experiences and worldview. All traditional cultures continually innovate within the context of their traditions, which are enriched and broadened in the process. However under the conditions that prevail in the modern commoditized growth economy, traditions that have the least commodity potential or compete in any way with the cultural hegemony of commercialization are threatened by innovation. Under the influence of commoditization the course of innovation leads primarily toward products and product improvements. Products gradually improve and temporary technical limitations are overcome, for example in electronic information storage, and entirely new lines of goods or services, such as the personal computer and the software and infrastructure that supports it, are created. It is generally assumed that innovation is positive, that the "guiding hand" of consumer preference naturally steers the process of invention toward ever-increasing quality and affordability.

This is true — within the parameters of commoditization. Those goods, however, that are least commoditizable, innovation leaves behind. There is no comparable force of continuous improvement for those things that are most culture and place-bound. These are exactly those aspects of culture most associated with what we call tradition. Traditional medicines, farming methods, native cuisine, dress, stories and music are all highly particularized, tied to a particular culture and methods that are passed down person to person. The particularized is always outcompeted by the standardized in a commoditized growth economy. When any culture is incorporated into the world economy, what is adopted and developed are those aspects of the tradition that are most commoditizable, like clothing and housing styles, food items, music, certain herbal treatments. These will tend to survive and even spread, often as specialty items for the wealthy. What is not commoditizable is discarded along the way, and this usually constitutes the most important, and certainly the most particular, aspects of a way of life.

Throughout its history every traditional culture also innovates, develops tools and techniques that improve results over time, and continues on a path that is self-reinforcing for that culture. However once commoditization becomes the leading driver of innovation and reaches a certain point where it operates to privilege those innovations that break the bonds of relationships within community and with the land, then innovation itself becomes the destroyer of tradition and culture. Given that traditions have evolved along with the climate and soils of a particular ecosystem (or in the case of hunter-gatherers or nomadic cultures, their particular range), these traditions store the memory of how to live in a place. The ecological consequences of losing that memory are significant. More frequent than the outright loss of traditional knowledge is the underdevelopment of that knowledge as development efforts get siphoned off into commoditized directions (more about this later). As a result, traditional knowledge may not have the opportunity to innovate and evolve as population grows and climates change, leaving the traditional ways inadequate to the times. Innovation is often correctly seen as the driving force of social evolution. Innovations are the social and conscious equivalents of genetic mutations which, if successful, will then survive and reproduce, shaping culture over time. If the only innovations that thrive are associated with commercial opportunities, then all cultures become increasingly commoditized and increasingly alike.

2.3.14 Simplification

Development as distorted by commoditization increasingly simplifies human life in the aggregate, even as it is experienced by the modern individual as becoming more complicated. As the private automobile outcompetes public transportation for development resources, quality public transportation grows scarcer, increasing the need for personal vehicles and thereby further underdeveloping public transportation. The experience of life growing increasingly difficult results from the increasing dependence of modern people. As more and more of the activities of daily living become dependent on commodities, the need for cash income grows. The quest for money becomes all-consuming, forcing people to sell more of the only thing they have to sell, their labor. As a result there is steadily decreasing amounts of time for noncommercial activities.

We mistakenly associate this shrinking availability of time with complexity. However, the actual processes involved in using the convenient products is considerably less complex than earlier, less commoditized versions of the same product or service. The modern washing machine takes far less attention and skill than earlier washtubs. Microwave cooking is simpler than stovetop. In general, our relationships with each other and with our material world grows ever simpler as the forces of commoditization operate. It is not that our lives grow more complex, but that the type of complexity changes.

2.3.15 Time frame of benefits

In general, benefits that accrue to investments of time, attention, and
resources toward community building, increasing knowledge of ecological
systems, and other activities associated with noncommercial products and
services tend to pay returns slowly over long periods of time, when they do
pay at all. Commoditization affects the world of finance and investment
much the same as it does everything else; money or other financial instruments that can serve as means of exchange, evolve in the commoditized
direction, increasingly mobile, increasingly detached from any real world
acts of production (this will be discussed in greater detail in Chapter 3). As
a result, investment capital is increasingly free to move toward the highest
and quickest returns. This in turn accelerates the process of commoditization,
as capital increasingly moves toward those investments with the greatest
commodity potential, a classic example of positive feedback accelerating the
whole process we are describing.

2.3.16 Sufficiency vs. efficiency

The goal of efficiency is to maximize output while minimizing input.
Standard measures of economic efficiency are not concerned with the
social or environmental benefits or damages associated with the output
being measured. There are no ways to compare these benefits or damages
with those provided by the noncommodity goods and services the products
may have replaced. If rather than the generally accepted measures of
economic efficiency we adopted as goals and indicators of human welfare
measures of sufficiency — optimal service for minimal expenditure of
material and energy — we might be in a much better position to compare
the benefits of HCP vs. LCP goods and services rather than merely comparing different competing commodities. In addition, sufficiency goals
might shift public policy toward providing the best quality of life with
the least amount of energy and materials expended: the necessary prerequisite for sustainable development.

2.3.17 Commoditization and GNP

The Gross National Product — consumer expenditures plus gross investment
outlays plus government expenditures on goods and services in an economy
— has gradually become the simplest aggregate measure of the size of a
nation's economy. In public policy deliberations it also became a surrogate
measure for the material well-being of the nation's citizenry. As such, it also
became a measurable target toward which to point national and international
economic policy. Yet even one of the economists who first formulated the
methods for calculating GNP, Nobel laureate Simon Kuznets, quickly grew
wary of the misuse of such gross statistics to determine public policy or
measure economic progress. "Distinctions must be kept in mind between

quantity and quality of growth, between its costs and return, and between the short and the long run,"[21] Kuznets wrote in 1962. It is a distinction that has gone largely unmade when gross product statistics are routinely used.

The problem with GNP is not the measure itself but rather the way it has come to be used as an indicator of the success or failure of economic policy.[22] GNP as an index measuring the size of the economy is as useful as air temperature in Fahrenheit (or Celsius) is for measuring the degree of heat: useful, but incomplete. Air temperature may not tell me all I need to know about my level of comfort, which is also affected by humidity, wind speed, surface temperature, and other factors, but it does provide useful information.

As the date of the Kuznets article shows, the misuse of GNP has been questioned almost since the beginning. Yet this critique has had almost no impact on actual economic policy-making nationally or internationally. This is understandable in light of the analysis presented here. GNP is a near-perfect measure of commoditization. Its use invariably reinforces the logic of commoditization, while commoditization feeds the logic of GNP. When grandparents provide voluntary babysitting, nothing registers in GNP; when the kids are sent off to a childcare center, a blip is noted in the GNP screen. Barbie adds to GNP, snow angels simply melt away. When crime rises in a city and people purchase increased security and alarm services, and replace goods that are stolen and hire additional police, all of these purchases are added to GNP. When you are well you add little to the medical goods and services category of GNP, but get sick and your tests and pills do their part for economic growth. When a hurricane blasts southern Florida, the repair and recovery contributes, and when an oil spill slicks Alaska, clean-up registers as a blip in GNP. The clean-up of toxic sites in the U.S. has cost billions of dollars and added those billions to GNP. First the products of manufacturing add to GNP and then the clean up of any wastes associated with that manufacturing process is also added. As a result, pollution-generating industries end up being counted twice. Unlike basic business accounting, depreciation is not subtracted from income, and the depreciation and loss of natural resources fails to turn up in the calculations. Not only does a rise in GNP fail to adequately reflect improving quality of life, it can do just the opposite by tallying all the "regrettable necessities," goods and services people buy in response to their declining quality of life or lack of personal and family time. It confuses ecological and social loss with economic gain.

In the early 1970s, economists Nordhaus and Tobin[23] responded to the increasing concern that GNP was not a good measure of economic welfare. They modified GNP to account for many of the early concerns to arrive at the Measure of Economic Welfare (MEW). A comparison of MEW with GNP from the 1930s to the early 1960s suggested the two were positively correlated. Daly and Cobb[24] looked at the time series for the first and second half of the years covered. In the first half the two measures were nearly perfectly correlated; in the second they were not, suggesting the relationship was weakening even over this early time period. Since then, evidence suggests dramatic weakening.

Daly and Cobb[25] describe the GNP modifications necessary to arrive at their Index of Sustainable Economic Welfare (ISEW), which deducted for pollution or depletion of natural resources. In addition, following standard economic understanding of the reduced value of each addition of something as it grows toward satiety and then extravagance, it counts a dollar added to the stock of the wealthy as less valuable than a dollar to the poor. Daly and Cobb compare a time series of GNP with ISEW[26] to demonstrate that the ISEW has leveled off despite a continued increase in GNP. Costanza et al.[27] compare ISEW with GNP across several different countries; again, the two are positively correlated until the early 1970s and then begin to diverge — dramatically in some countries.

The San Francisco group Redefining Progress uses a Genuine Progress Indicator (GPI), which follows Daly and Cobb in their deductions for depreciation of natural capital but also makes additions for uncounted, noncommoditized services such as the value of household work and the provision of ecological services from "undeveloped" natural areas. Figure 2.3 summarizes the trend in GPI in relation to GNP.

None of the proposed alternatives or corrections to GNP is likely to matter until public policies are deliberately designed to counter the effects of commoditization. Once that occurs, indices such as the GPI or ISEW can be used to the track gains in public welfare as measures of successes or failures of those public policies. Indicators in the absence of common goals carry little meaning. By understanding how commoditization works, we can identify those aspects of life that remain underdeveloped. Our goals can then be formulated in terms of true development, in which the quality of our lives gets progressively better, rather than just quantitatively bigger in terms of the amount of the Earth's materials and energy we manage to transform into waste.

All of this is logically consistent with commoditization. What counts is not the service provided, but the movement of more and more of economic life into the market. As a result, growth in GNP invariably reflects a shift in the character of a society away from the qualities and characteristics associated with noncommerical goods (see Table 2.2) toward the qualities and characteristics that make something a commodity. To the extent that this leads to an imbalance favoring one side and impoverishing the other, our quality of life will be in decline. Most of the measures of public welfare created and promoted as alternatives to GNP measure a slowing of general improvement and in some cases a decline in quality of life in industrial societies beginning in the 1970s, despite sharp rises in GNP from then until the later 1990s.

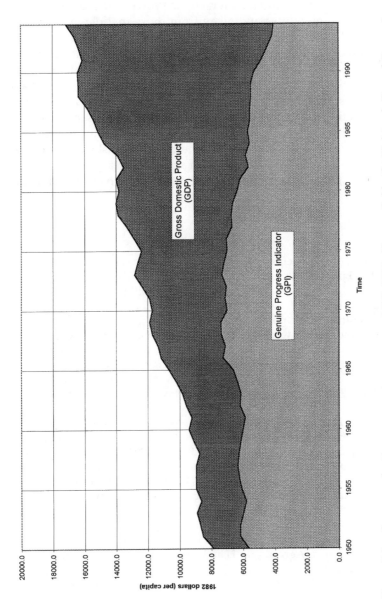

Figure 2.3 Gross production *vs.* genuine progress, 1950–1995. (*Source:* Cobb, C., Halstead, T., and Rowe, J., Redefining Progress, 1995.)

2.4   Conclusion

Commoditization is a selection pressure that favors those goods and services that are fit for serving as commodities in the exchange economy. As a result, the social world of human interactions evolves toward the qualities that characterize commodities and away from those that defy commoditization. Table 2.2 lists the qualities most associated with goods and services with high commodity potential as well as the counterpart qualities associated with noncommercial goods. A look over the differences between these two sets of qualities gives us a good summary of what we have gained and what we have lost by allowing commoditization to distort our economies. In effect commoditization favors those things that have been separated from their web of concrete ecological and social relationships and placed in a new context of increasingly abstract commercial relationships. Chapters 3 and 4 look more closely at the process of commoditization and provide examples of how it distorts economic development in each of several sectors of the economy.

Notes

1. Consider these numbers in light of the fact that the combined wealth of the richest 225 individuals in the world is $1 trillion dollars. This means that an annual levy of 4% on the wealthiest 225 of the wealthy would suffice to meet all the basic needs of the world's poor. This $1 trillion is equal to the combined income of the world's poorest 2.5 billion people.
2. For a picture of the working conditions of some of the toy industry, see Williamson, H., China's toy industry tinderbox, *Multinational Monitor*, 14(9), September 1994, 24. For a look at the industry, see Pecora, N.O., *The Business of Children's Entertainment*, Guilford Press, New York, 1998. See also Cross, G., *Kid's Stuff: Toys and the Changing World of American Childhood*, Harvard University Press, Cambridge, MA, 1997.
3. Stevens, T., Playing to win, *Industry Week*, November 3, 1997.
4. Editors of the Ecologist, *Whose Common Future? Reclaiming the Commons: the Ecologist*, New Society Publishers, Philadelphia, 1993.
5. Klotz, C., Fuglie, K., and Pray, C., *Private-Sector Agricultural Research Expenditures in the United States, 1960–92*, U.S. Department of Agriculture, Natural Resources and Environment Division, Economic Research Service Staff Paper No. AGES9525, October 1995.
6. U.S. Department of Agriculture, *Agricultural Resources and Environmental Indicators, 1996–97*, Economic Research Service, Natural Resources and Environment Division, Agricultural Handbook No. 712, Washington, D.C., July 1997, pp. 246–247.
7. See Aguilera-Klink, F., Some notes on the misuse of classic writings in economics on the subject of common property, *Ecological Economics*, 9, 1994, 221–228. Also see Berkes, F., The benefits of the commons, *Nature*, 340, July 13, 1989, 91–93. Also see Ostrom, E., *Governing the Commons: The Evolution of Institutions for Collective Action*, Cambridge University Press, New York, 1990.
8. Gates, J. *The Ownership Solution*, Perseus Books, New York, 1998, p. 189.

9. Internet shopping may yet establish the economic limits of customer mobility. With instant access to price information from multiple competing sources, consumers can easily do comparative shopping for the best price. As a result, prices will likely be driven down to near manufacturing costs. This will further strain the economic viability of local merchants, threatening employment in the retail industry, which has been one of the strongest job producers in the modern commoditized economy.

10. Pimentel, D., Stachow, U., Takacs, D.A., Brubaker, H.W., Dumas, A.R., Meaney, J.J., O'Neal, J.A.S., Onsi, D.E., and Corzilius, D.B., Conserving biological diversity in agricultural/forestry systems, *Bioscience*, 42(5), May 1992, 354.

11. Weatherford, J.M., *Indian Givers: How the Indians of the Americas Transformed the World*, 1st ed., Crown Publishers, New York, 1988, p. 62.

12. Horton, D.E, Lessons from the Mantaro Valley project, Peru, in *Let Farmers Judge: Experiences in Assessing the Sustainability of Agriculture*, Hiemstra, W., Reijntjes, C., and van der Werf, E. (Eds.), Intermediate Technology Publications, London, 1992.

13. The Wonder Years, changes in franchising and the Franchise 500 since 1979, *Entrepreneur*, 27(1), 180, January 1999, http://web3.infotrac.galegroup.com/itw/session/526/624/11406033w5/, 4/15/99.

14. Lovins, A. and Browning, W.D., Negawatts for buildings, *Urban Land*, 51(7), July 1992.

15. A good summary of system thinking can be found in Peet, J., *Energy and the Ecological Economics of Sustainability*, Island Press, Washington, D.C., 1992, Chapter 5, pp. 72–81.

16. Kohn, A., *No Contest: The Case Against Competition* (rev. ed.), Houghton Mifflin, New York, 1992, p. 182.

17. Hall, C.A.S., Cleveland, C.J., and Kaufmann, R., *Energy and Resource Quality: The Ecology of the Economic Process*, University Press of Colorado, Niwot, CO, 1992.

18. Hall, C.A.S. and Hall, M.H.P., Issues in global agriculture and environmental sustainability: the efficiency of land and energy use in tropical economies and agriculture, *Agriculture, Ecosystems and Environment*, 46, 1993, pp. 1–30.

19. Hall, C.A.S. and Hall, M.H.P. (1993).

20. Norgaard, R., *Development Betrayed: The End of Progress and a Coevolutionary Revisioning of the Future*, Routledge, London, 1994, p. 27.

21. Kuznets, S.S., How to judge quality, *The New Republic*, 147, October 20, 1962, 29.

22. Cobb, C., Halstead, T., and Rowe, J., If the GDP is up, why is America down? *The Atlantic Monthly*, 276(4), 59, 1995.

23. Nordhaus, W. and Tobin J., Is growth obsolete? *Economic Growth*, National Bureau of Economic Research General Series, no. 96E, Columbia University Press, New York, 1972.

24. Daly, H.E. and Cobb J.B., Jr., *For the Common Good: Redirecting the Economy toward Community, the Environment and a Sustainable Future*, Beacon Press, Boston, 1989.

25. Daly and Cobb (1989), p. 401.

26. Daly and Cobb (1989), p. 420.

27. Costanza, R., Cumberland, J., Daly, H., Goodland, R., and Norgaard, R., *An Introduction to Ecological Economics*, St. Lucie Press, Boca Raton, FL, 1997, p. 136.

chapter three

Evolution, systems, and commoditization

Contents

3.1   Introduction

There is no aspect of human life that has not to some extent been commoditized, transforming intrinsic and communal values into commercial values. The pressures that privilege commercial goods over common goods involve a gradual selection process, a "survival of the fittest" where what is fit is by definition what is potentially marketable. The selection process for commoditization is thus circular and self-reproducing.

The differences between high and low commodity potential are matters of degree. Goods and services more or less have the qualities of a commodity: they are more or less mobile, alienable, standardizable, and so on. Goods with low commodity potential are less alienable (more communal), less mobile (attached to local ecosystems or local culture), less marketable (being communal and attached), and less standardizable. Most things are some of each. Even noncommodities such as "friendship" have their commoditized service version in "psychic friends networks" and personal ads. I speak here of "commodity potential" as a relative distinction between different goods and services. Virtually everything has some commodity potential. Commoditization as a selection pressure encourages and coaxes the most commodity-like qualities in everything.

The characteristics of products, like those of plant and animal species, emerge through a gradual evolutionary process. Rarely is one species selected over another, but instead one minor qualitative difference may enhance survivability and reproductive potential in a particular environment, and so those individuals of a species that carry the optimal degree of that quality are more likely to contribute genetic material to their descendants, the future participants in that particular web of ecosystem relationships. Their offspring are statistically more likely to also carry the superior quality, and on and on.

The selection process that results in directing investment toward commoditizable goods and services is not necessarily more powerful than the other factors, such as physical suitability and consumer choice, that also determine which types of goods and services will be developed. Just because something carries the qualities of a commodity in the sense I have been describing does not mean it will sell. However, commoditization plays enough of a role so that over time, more and more HCP goods outcompete LCP goods. Commodities that are not also intelligently designed and attractive to consumers, that are not excellent solutions in their own right, may not survive in any case.

The great success of modern economies is the incredible array of material goods and services available for sale to satisfy human needs. The feedback loops built into the structure of the current economy give positive rewards for investing time, attention, and resources into producing the most commoditizable of goods and services; information is fed back to those who control investment that allocating these resources toward commodities will yield far better returns than equal investments in noncommodities. Some have argued that the way to fix the distortions of economic development is to distribute control over investments to more stakeholders by capitalizing consumers, workers, and residents of the communities in which production or resource extraction occurs. This could be achieved by various schemes for what is sometimes called democratizing capitalism.[1] Although such reforms are certainly warranted and would go a long way toward correcting the gross inequalities inherent in the present patterns of ownership of productive resources, it would not likely address the systematic bias of commoditization and the fetish of economic growth.

Successful economic growth *is* the success of commoditization. But as Herman Daly and others have pointed out, there is a big difference between *growth* and *development*. Growth is concerned only with the expanding volume of goods and services moving in an economy. Development, by contrast, should include many other considerations that affect personal well-being: state of physical and mental health, literacy, family life, etc. In our terms, true development also includes many of the aspects of life that are least commoditizable. Not only does the obsession with economic growth exclude considerations of noneconomic factors, but it makes it more difficult to achieve economic and social development in the LCP sectors which remain as a result relatively underdeveloped.

What we need is balanced development, which facilitates commercial, cultural, and personal development alike. Such development must therefore include the discovery and development of innovative solutions to satisfying human needs with little or no commodity potential. The systemic nature of commoditization makes it unlikely that countervailing forces that will develop LCP goods and services will emerge without direct policy intervention. However, in order to invent and apply effective policies to stimulate LCP development we must first understand the evolutionary dynamics of commoditization itself. This chapter will argue that commoditization acts as a Darwinian selection pressure that, by "selecting" those species of goods and services that utilize more energy and raw materials and create more waste, drives the development of the entire economy toward ever-increasing mobilization of energy and materials, pollution, and waste.

We begin by looking at products to see how commoditization serves as a selection pressure, an evolutionary force that favors certain qualities in goods and services over others. We then shift focus and look at the system level in which products, producers, and consumers interact. Systems evolve both as their makeup changes through the evolutionary dynamics of their component parts and through dynamics that operate only at the system level. Then we look at how the process of commoditization operates both at the level of individual goods and services and at the level of the economic system as a whole to affect several economic sectors including agriculture, health care, environmental protection, science and academia, and social movements.

3.2 Commoditization and evolutionary theory

At the core of much evolutionary thinking lies the notion of competition among organisms for limited resources. In nature, the limited resources are sunlight, prey, habitat, or other sources of energy and nutrients required for physical maintenance, daily activities, growth, and reproduction. Certain types of organisms will exploit available resources fastest or will develop strategies to keep others away from the resource. The physical features of the environment create the conditions for evolution. In effect, the physical conditions present a system of incentives for certain qualities and features to emerge and disincentives that discourage others.

In an economy, the limited resources include energy and nutrients (raw materials) but also human time, attention, and money — or investment (each reflecting allocations of energy and nutrients). The most important added ingredient in economic and social development that makes it distinct from "natural evolution" is the creativity and intelligence of human beings focused on finding solutions for problems related to obtaining the material means to survive and thrive. As I have noted, for any human need (for nutrition, for healing, for transportation, for child's play, for leisure) there is a range of options for satisfying it. These can be considered competing forms, or species, of goods and services; they are competing for investments of time, attention, and resources. New ideas, new ways of doing things, new information are analogous to new variations being introduced in an established environment. New forms (new products or services) then compete to obtain what they need to survive, develop, and reproduce.

In the natural world, each living thing continually interacts with its environment and the particular trophic, parasitic, or symbiotic relationships it maintains with other living things to optimize its chances of getting food (material and energy) and passing on its genetic information through reproduction. Wherever there are opportunities there are niches for life of one form or another to fill. Out of this self- and gene-maximizing behavior, ecosystems emerge as a structured flow of energy and materials in a context of particular geologic, chemical, and physical conditions. Ecosystems and the creatures that inhabit them coevolve.

The human economy and all the species of goods and services that exist within it also coevolve. To understand how, we can look to Darwinian principles of natural evolution as an analog for all forms of evolution. Evolution always involves at least three factors: initial conditions, variation or modification (i.e., generation of diversity), and selection. In natural systems the experimentation consists of the ways in which genetic material is randomly rearranged through accidental mutations. The resulting new features from chance genetic change that are most fit for survival under the particular environment in which they occur are gradually selected. Rather than competition between species, we can understand the process as a cooperative system producing the optimal forms of life for any given conditions. The process is never static. Random events change environmental conditions, which then change the parameters of what is fit and reconfigure the flows of energy and nutrients and the various possibilities for drawing on those flows (niches). The relationship between evolutionary dynamics and the system within which they operate is thus interdependent. The activities of all the life-forms together gradually, and sometimes suddenly, change the physical conditions in ways that also transform the definition of what is fit in an ecosystem. In this mutually creative and destructive way, life-forms and their environment coevolve.

Niche is a critical concept in both ecology and economics. In nature, a niche is a distinct arrangement of a particular set of limited available nutrients and energy.[2] Living organisms mobilize the available energy and nutrients and draw upon them for self-maintenance and reproduction, and if comparatively successful, a species is said to occupy a particular niche. Niches are as diverse as the geophysical environment itself. The more varied the niches, the greater the diversity, and vice versa. Certain species specialize in particular niches and can survive in no others. Generalists can thrive in a range of conditions and can come to dominate many niches at the expense of the specialists. Extremes of variability and stochastic processes (random, unpredictable) tend to shake up the environment enough so that new niches are constantly created and the development process starts anew. Though it was long thought that all ecosystems have final climax stages that they can maintain indefinitely if not disturbed, we now know that this is not the case. Indeed, we recognize disturbances as normal parts of ecosystem function and evolution. There is no final climax state in nature. Unpredictability and change are ubiquitous and favor the flowering of specialists and with them, diversity.

The process of change never ends and the potential for biological innovation through mutation and drift is virtually limitless, but the actual manifestations of evolutionary change are not infinite. They are limited by the laws of physics and chemistry, which determine the range of possible survivors. There are propensities for certain kinds of structure and behavioral strategies that derive from the natural laws which determine the most efficient and most likely pathways for the flows of energy and materials. *Although the conceptual possibilities of experimentation, innovation, and invention are unlimited, what actually has a chance of surviving is determined by the structure of energy and material flows.*

The human economy is also a structure of energy and material flows that determines what species of goods and services survive. Commoditization directs the energy and material flows as well as the human attention that determines those flows toward certain species of goods and services with the quality of commodities. If the structure of the economy limits our capacity for invention and innovation in ways that are destructive to us, as I believe it does, then the structure must be changed. The economy is subject to our deliberate manipulations and plans. Thus, if due to the structure of the economy species of goods and services are selected that are most costly in terms of mobilizing energy and materials, then we have an extremely difficult task in limiting these environmentally expensive products. However, if we deliberately modify the structure of the economy so that less environmentally destructive flows of energy and material are encouraged, then more environmentally desirable products will be produced. Changing the economic structure means changing the selection pressures that now privilege HCP goods and services.

3.3   The natural selection of commodities

The pressures that select for commodities or consumer goods over common goods are not particularly mysterious. They are aspects of the pattern of human development, an extension or an enlargement of the pattern of biological development, operating through mutation (innovation, invention, serendipity) and selection. Certainly, economic selection is different than natural selection in that its path is in some sense guided by deliberate and intelligent effort — it is nonrandom. Things are tried in anticipation of fitness, of profitability. The selection pressures that determine the flow of money and energy and that attract human attention are, in the modern industrial economy, exactly the forces of commoditization I have been describing. Some mutations are rewarded with investments in R&D and others are not, or some are rewarded grandly and others stingily. The reward process is skewed in the direction of commodities and potential commodities.

New ideas and possibilities are numerous, although they are significantly constrained by history. Only those that are "practical" are likely to survive. What is "practical" is historically determined by what has succeeded in the past and has constructed the present. The way the past constructs the present is a key factor in commoditization. Thus another determinant of survivability, both in ecology and economics, is the degree to which an innovation restructures the environment in which the new species or product now exists. This happens whenever changes in the environment create multiple possibilities for stable structure (or multiple potential equilibrium states), and when efficiencies of standardization require that one structure emerge as the most preferable. Consider the hypothetical example of drivers free to choose to drive on either side of the road. Assuming that the choice in the beginning is random, there are no obvious pay-offs for selecting either side of the road. However, starting from random selection, depending on the choices of the first few drivers, a preference for one side or the other will gradually appear. Once it does, with increasing seriousness one side or the other will quickly become safer until being in the minority is suicidal. Despite no initial difference in the benefits of either side, the rational choice quickly becomes obvious.

There are many examples of this kind of *path dependency*. One example is the story of how the letters of the standard computer keyboard (and before that, the typewriter) came to be laid in the familiar QWERTY pattern. Because the original typewriters were prone to frequent jamming, commonly paired letters were separated in order to deliberately slow the typist down so that these paired keys would not be struck simultaneously.[3] Studies have shown that better layouts exist that would make typing faster and easier to learn. But by the time anyone attempted to improve key layout, QWERTY was thoroughly entrenched. For anyone considering making the change to a new keyboard layout, the retraining and reequipping costs were greater (or at least appeared to be) than the efficiency that could be gained through change. Once QWERTY became the standard, innovation in keyboard layout was stifled. Similarly, the VHS videotape standard outcompeted BETA, despite some apparent advantages of the BETA technology. It was simply too expensive to change to another system once most people had purchased VHS machines, and a generation of technology had to pass before another innovation could emerge, such as the DVD.

This phenomenon, through which it is possible for an inferior technology or design to survive, is most likely to occur when the product is part of a system of interdependent components, each designed to work with the other in providing a service. QWERTY keyboards were part of a system of document production that involved significant investments in machinery and worker training for which standards, any standards, were far more economically preferable than none at all.

Likewise, in the early development of personal motorized vehicles, steam, electric and gas engines all "competed" to become the standard in a majority of cars. The engines were just one part of a personal transportation system that would require a transportation infrastructure, including millions of refueling stations. What began as a support system for gasoline engine refueling and repair became a part of the benefit package associated with gasoline engines. After time, gas engines predominated to the exclusion of alternative options, even if the other options might have been more efficient or less polluting. Once a species of good or service occupies a particular economic niche, it is far more difficult for an alternative solution to unseat the present occupant in that niche. The present occupant extensively modifies the environment (the economy) in which newcomers might compete and usually modifies the environment to maximize the amount of energy and materials flowing in its direction. Thus, the automobile mobilizes a tremendous flow of energy and materials in the form of gas stations, repair shops, dealership, highways, and on and on. The new environment places alternative transportation, be it solar cars or mass transit, in a much more difficult competitive position. As the economy develops in a more and more commoditized direction, further commoditization gets locked in. The more certain goods and services develop and the more they are used, the more is learned about them and the more they improve. The types of goods and services that are treated preferentially in the allocation of investment resources become more highly developed, and the less commoditized alternatives appear "backward" in comparison.

The present economic environment consists of what has succeeded, or survived, in the past. Commodities, because they have received the bulk of R&D in the past, are in the present far more developed than competing noncommodities. Just as organisms that are most proficient at utilizing energy have the most likelihood of survival, so too do those goods in an economy that are most proficient at utilizing money, in other words commodities. Over time, this differential in development increases. Hence the acceleration and expansion of commoditization as the process gains momentum. The selection pressures I have described as commoditization function in the present, but have their greatest impact as forces in history that limit present and future options.

Many are familiar with the free market sense of this process. Many potential products are offered that compete in the market. A selection process ensues in which buyers and sellers allow some products to "survive," meaning they stimulate enough purchases at the necessary price to turn a profit. Many factors go into this selection process, but the most important is how successfully the product fits a particular market niche, or how well it creates a new niche.

We all understand that in practice the "competition" is only approximately "fair" and that money (for advertising, promotion, subsidies, packaging, or the ability to wait out early market losses) begets more money and that the "winners" do not necessarily reflect the "best product" but only the one that has survived because investors realized a net gain. In spite of imperfect markets, however, the selection process of consumer choice generally works well enough to push commodities to evolve toward greater levels of consumer satisfaction. This is the familiar selection process of the "invisible hand." This is the surface level "competition," through which commodities compete with each other to occupy market niches.[4] This is not the process of selection I am referring to in discussing commoditization pressures, because the competition for market niches is only open to commercial goods in the first place.

Other related but more fundamental prior selection pressures subtly determine the structure of the competitive system itself and apply considerable preference for high commodity potential over low commodity potential options for need satisfaction. The "invisible hand" as a selection process selects only those commodities that fill market niches. But a niche is a particular set of possibilities for obtaining and exploiting material and energy resources. The allocation and distribution of material, energy, and human intelligence determine what market niches become available. This fundamental allocation and distribution is in turn subject to selection pressures that favor certain patterns of material, energy, and attention over others. In a market economy the ones that are favored are, of course, the ones which function best in a market.

Here lies the fundamental positive feedback mechanism that shapes the pattern of production and consumption. Economic niches open much more fully and often for high commodity potential commercial goods than for low commodity potential goods or noncommerical and common goods. Once opened, niches are then filled by an array of high commodity potential goods selected by the interactions between buyers and sellers. Commoditization pressures thus operate at a more fundamental level than the level of the "invisible hand" of capitalism. They operate primarily in determining the flows of investment of materials, energy, and attention, and these flows define what is meant by a "niche."

There are at least three sets of constraints that together act to select economic survivors: consumer choice, physical limits including resource availability, and the ability to attract research and development investment. On the surface, it may seem that the category of "investment attractiveness" is not independent of the first two, because investments are attracted to what will meet customer needs, to what is physically plausible and can be accomplished efficiently with available resources. But there is an extremely important, if patently obvious, component to R&D investment decisions that creates a systematic selection bias for HCP commodities over LCP commodities and noncommodities: that the end product of the R&D investments not only must be physically plausible and meet an actual (or potential) customer need but also should be as broadly and profitably marketable as possible. The obviousness of this economic fact is perhaps why little attention has been paid to its ecological consequences. Investment functions as a selection process that drives the evolution of the economy toward HCP solutions, which are solutions with the greatest environmental impact.

These three constraints can be illustrated in the design of a car. Physical laws determine the range of design options and as a result most cars look fundamentally alike: height, width, wheel span, steering mechanisms, etc. Designs that veer too far from the physically optimal pay the price in higher fuel demands, safety risks, or other inefficiencies and are selected out of the production process. Consumer choice determines most of the variability within the range of what physical laws determine is practical. These choices are the result of the interplay of options and motivations, including disposable income, status seeking, comfort, and practical considerations such as size of family, the purposes for which the vehicle would be put to use, etc. The combined result of these factors leads to a range of available vehicles and features on the market that represent the optimal balance of possible customer satisfactions. Neither physical limits nor consumer choices, however, account for the selection of automobile transport over other, forms of transportation such as adequate public transport, redesigned cityscapes that minimize transport needs, etc.

The fact that high commodity potential forms of need satisfaction such as automobiles receive orders of magnitude larger investments in research and development than low commodity potential alternatives for the same economic niche means that personal transport receives the means of survival and reproduction while the alternatives suffer deprivation of materials, energy, and attention — in other words, investment.

Once the market has been structured to provide incentives for high over low commodity potential choices, it is very difficult to change it back. For instance, in the 1930s a conglomerate of American car, oil, and tire companies bought up and promptly dissolved over 100 street car lines in American cities.[5] Add to that federal subsidies for highway construction and the market is severely biased toward cars rather than public transportation. Nonprofit and public investments tend to be skewed by the same forces. Once the pattern of need satisfaction is set in place through large-scale private investment in HCP-based needs satisfaction, public expenditures appropriately follow the logic of investing in the greatest good for the greatest number. Once private automobiles and their related goods receive the bulk of private investment, public investments in common goods such as roads and support systems such as licensing and registration systems become necessary. A positive feedback is set up,[6] better roads make cars more attractive, more cars require more roads and so on. At the same time, more drivers make for fewer public transit riders, and the economics of public transportation become increasingly marginal until it survives only in densely populated urban centers. No amount of trying to convince people that public transportation is more environmentally sound can change this dynamic, because changes in values alone cannot create the conditions necessary to generate investment and create a market.

This phenomenon, the selection of high over low commodity potential solutions to personal and social needs can be observed in sector after sector. It is a self-reinforcing, positive-feedback mechanism. The key selector is investments in research and development. Since high commodity intensity solutions receive by far the greater amount of R&D, they invariably appear to be more advanced and competitive. It is then logically compelling to utilize these solutions and apparently irresponsible to suggest the adoption of lesser developed alternatives. Communal solutions therefore always appear less capable of meeting human needs.

The comparative advantage of high commodity potential goods is exacerbated by the tendency to think of development as progress. The progress achieved by commoditization-driven economic development is juxtaposed to a lack of development (underdevelopment). Possible courses of development not taken are not considered. It is impossible to compare present conditions with what might have been if alternative paths had been taken. The comparison between present development and untried past alternatives is always an unfair comparison.

Anyone bringing a new product to market has to have a cheaper or better product to elbow his way into a market against well-established competition. You can't do merely as well as established practice, for otherwise no one would be encouraged to switch over to your product. In the end, the more efficient and best designed will survive. There are two sets of characteristics in the making of a new and better product. First it must meet some need in a cost-efficient manner and second it must be marketable — it must serve as a commercial product, a commodity. What happens when the best educational practice, the best health maintenance, the best methods for growing food and other best practices involve knowledge or practices that are not best suited for the market? The invisible hand of the market will stimulate development, but such development will be only in the direction of commoditization.

3.4   Commoditization and systems

We have looked at development so far strictly as analogous to themes in evolutionary biology. Theoretical ecology and systems dynamics also can help us to understand how commoditization works.[7] Commoditization in excess results over time in increased "brittleness" and decreased resilience of our economies and societies; it makes them more likely to break under stress. As more and more people grow increasingly dependent on commodities for their basic survival and fewer and fewer people can depend on mutual aid and community self-reliance (noncommoditized economic values), the dangers associated with economic disruption increase dramatically. If we understand that both economic system and ecosystems are systems of material and energy flows, we can then draw from the insights of theoretical ecology to help understand what's wrong with a commoditized economy. In Nature, an ecosystem gradually trims out its least ecologically efficient energy and material pathways and continually adds new connections, or niches, to take advantage of every possible nutrient and energy opportunity.

Each element in a system acting independently tries to maximize its access to the nutrients and energy it requires to survive and thrive. But it relies on system processes to maintain steady flows of those nutrients and energy. It cannot maximize its own short-term consumption of resources at the expense of the system without compromising its own longer-term survival. All creatures have evolved in the context of ecological systems both large and small and have evolved strategies that take advantage of particular opportunities in the available network of nutrient and energy flows. All systems, if they are to survive as systems, find a sustainable balance between individual maximization and collective optimization. Commoditization is in effect the individual maximization principle by which each economic actor attempts to maximize her or his portion of the flow of economic nutrients and energy (financial resources) regardless of the consequences for the system as a whole. Reducing more and more people's capacity for mutual aid and self-reliance and making them more dependent on a single global network of economic exchanges makes everyone's existence more tenuous. Unless commoditization is balanced by an optimization principle through which economic nutrients and energy are allocated in ways that make the whole system more resilient and sustainable, then the system itself becomes increasingly unstable.

Systems theory can help us to understand how unhampered commoditization threatens economic stability and sustainability. Systems are recognizable as systems because they are organized in patterns that create some enduring structure, which has properties that are different than the sum of its parts. These are often referred to as a system's *emergent* properties. The relationships between the parts of a system are the most important feature of any system.

The structure of an ecosystem consists of an intricate web of nutrient and energy exchanges where one organism's waste is another's food, one's respiration is the stuff of another's inspiration, and organisms are food for one another. Life constantly operates to maximize itself. The maximal amount of life in any ecosystem is determined by the amount of recycling and multiple cycling and the efficiency of each exchange. Efficiency in this context is measured as the percent of loss of energy and material from the exchange per unit of flow. This is *not* the same meaning given to the concept of efficiency in economics, which is concerned primarily with those monetary costs. Ecological efficiency means getting the most life out of the least possible inputs of nutrients and energy. It maximizes utilization and minimizes losses from the system. By comparison what is usually meant by *economic* efficiency is a measure of the quantity of goods and services produced per unit of consumer cost. Economic efficiency in this way *could* be made to approximate ecological efficiency if — and it is a very large *if* — economic costs represented a measure of nature's costs; in other words, if ecological values were represented in the system of economic exchanges. If this were true, then the economy would begin to evolve toward efficiency in terms of getting the most quality of life per unit of energy and nutrients (raw materials) used. Commoditization, by systematically detaching economic goods from their set of ecological relationships, makes this extremely difficult to do. This is why systems thinking is so hard to implement in a commoditized economy. Systems are about relationships. Commodities perform best as commodities when they are abstracted from or removed from their web of relationships.

In Nature, ecological efficiency is gained in two ways that *appear* to be incompatible: through specialization *and* consolidation. Each pathway in an ecosystem network can be understood as an ecological niche. Within each niche, over time, the least efficient species is outcompeted by the more efficient, and over time the tendency is toward fewer and fewer species per niche, or loss of diversity. At the same time, the elements of the network grow increasingly specialized to take advantage of every possible access point to the available nutrients, and energy and diversity increases. Maturity means both increased complexity and increased efficiency.

As efficiency increases, diversity at the level of niche decreases, but the number of niches increases. In early successional systems a few pioneer species adapted for widespread dispersal and rapid growth do all the work of mobilizing the available energy and materials rapidly but inefficiently. After limitations of available nutrients or sunlight set in, other more efficient species enter the scene. As long as growing conditions remain relatively stable, the more efficient, slower growing species begin to predominate. With an even longer period of stability, a wide variety of species will move into the many slightly different niches along gradients of nutrient, sunlight, and other growing conditions. There are many potential pathways of material and energy transfers from primary producers to carnivores each taking place simultaneously in a highly developed system. In this way, both diversity and efficiency increase with system development. Within each niche only the most efficient species dominate, leaving only traces of their competitors — but diversity, efficiency, and throughput within the system increase as niches are constantly subdivided with ecosystem stability. In this way the system maximizes its use of resources given the prevailing stable conditions. But it does so at the expense of resilience attributable to those species and flow pathways that were more competitive under different conditions. If the system is stable for so long, those species may disappear all together. With stress or perturbation the system is susceptible to the degree to which the resilience of the system has been pruned.

Efficiency can mean very different things at the level of the product and at the level of the system. Commoditization, as we saw earlier, encourages product-level thinking and planning rather than system-level attention. As a result, production efficiency at the level of the product is favored over system-level efficiency, even though the resulting distortions to the system ultimately threaten production efficiencies for everything. Product-level efficiencies ignore the impacts on the larger system of life support that includes environmental and social conditions in addition to conditions suitable for maximum production and consumption. The major environmental problems of our time are all system-level problems: global climate change, water scarcity, proliferation of toxic chemicals, stratospheric ozone depletion, global loss of biodiversity. They can only be addressed through system-level responses. These require a willingness to invest in goods and services that are important for the health and vigor of a society but cannot be readily exchanged.

At the level of the system, product-level inefficiencies (associated with noncommodity system elements such as those related to community infrastructure, governance capacity, mutual aid, and community self-reliance) benefit the system in at least three ways:

- They increase the number and variety of patterns of development and facilitate the recycling of energy and materials, thereby increasing the number and varieties of pathways to development (and therefore increase total development capacity).
- They direct resources to system maintenance.
- They promote redundancy for added stability and resilience.

In referring to similar processes in natural systems, Ulanowicz calls all of these aspects *system overhead*.[8] According to him, the aging and senescence of systems result from overdeveloping production efficiencies and underdeveloping overhead. What should be our economic goal is the "best" solution: optimal efficiency in commodity production *and* continued development of the overhead. If we allow highly commoditized goods to completely replace less commoditized means of achieving the same result, we will lose the ability of substituting the less commoditized alternatives in the event of some stress to the system — such as the environmental or social stress likely to come as the result of too much commoditization.

Commoditization acts in the economic system much the way that the propensity for efficiency of flows acts in an ecological system. New inputs into the system, in the form of investments, are captured in disproportionate amounts by the most efficient and most fluid of the potential exchanges. Investing in factories that mass-produce Barbie dolls is an efficient use of energy and material to get the most play value to the greatest number of children. To use the same amount of resources to improve systems of work and play in order to free more time for more direct child/adult play would send the same investment in countless, less efficient directions. As commoditization acts as a selection pressure at the level of individual goods, it plays out its evolutionary consequences, and resources are directed toward expanding the most economically efficient flows until they suck up nearly all the available energy and materials.

Efficient investments are favored for their ability to reduce overhead and increase production and mobility (liquidity). Consequently, fewer and fewer pathways exist for the less mobile, less efficient goods and services. As a result the economy grows simpler (in terms of the mix of HCP and LCP goods and services), more economically efficient, and more brittle. In nature there is always a check to this trend; some form of severe environmental stress overwhelms the simple, overdeveloped structure and sets it back to rebuilding connections, relationships, and structures in a multitude of experimental pathways. Thus, while short-term development is toward maximum efficiency (maximum production, minimum overhead), long-term evolution will favor optimal production and optimal reserves and redundancy. In the human economy we don't need to wait for some severe economic stress to check the overdevelopment of the highly commoditized parts of the economy. We can instead intervene with public policies specifically designed to provide countervailing pressures against commoditization to allow the economy to develop along more balanced and optimal pathways. We will consider these policies in more detail in Chapter 8.

3.5   Conclusion

Using concepts from evolutionary biology to understand the development of our economic system is not unproblematic. However, it offers more advantages than disadvantages. The means by which a human need or want is satisfied does indeed depend largely on a process of selection among various real or potential goods or services. The "environment" in which selection of goods and services takes place is a given economic system. Within any system that is self-sustaining and self-perpetuating, an evolutionary process is established that favors the continuation of that system. Characteristic of evolutionary processes in an economy are certain selection pressures that help determine which goods and services the economy will produce. Since it is the energy and materials associated with these selected goods and services that ultimately are responsible for the economy's impact on the environment, it is useful to understand and be able to moderate this selection process in order to minimize its impact. Consequently, it is appropriate to draw upon the concepts of both evolutionary biology and system theory to make sense of the process of commoditization.

Notes

1. For the most thorough and persuasive presentation of the path to democratic capitalism, see Gates, J., *The Ownership Solution: Toward a Stakeholder Capitalism for the 21st Century,* Penguin, London, 1999.
2. Hutchinson, G.E., *Population Studies: Animal Ecology and Demography,* Cold Spring Harbor Symposia on Quantitative Biology (concluding remarks) 22: 415–427, 1957; Elton, C., *The Pattern of Animal Communities,* Methuen, London, 1966.
3. Olson, D.W. and Jasinski, L., Keyboard Efficiency, *Byte,* February 1986, pp. 241–244.
4. Boulding, K.E., *Ecodynamics: A New Theory of Societal Evolution,* Sage Publications, Beverly Hills, CA, 1978. Product introductions are part of what Kenneth Boulding has called the "mutational structures of society." They provide the continuous flow of innovation upon which consumer selection acts and through which technology and material culture evolves.
5. Kuntsler, J.H. *The Geography of Nowhere: The Rise and Decline of America's Man-Made Landscape,* Simon & Schuster, New York, 1993.
6. Boulding (1978).

7. Ulanowicz, R., *Ecology, The Ascendant Perspective*. Columbia University Press, New York, 1997. For this section I rely largely on the work of Robert Ulanowicz and his theory of ascendance.
8. Ulanowicz (1997).

chapter four

Commoditization and the distortions of development

Contents

4.1 Introduction

Once a people's basic material needs are met, economic progress can be expected to result in the steady increase in the number of opportunities available to the population: larger variety of foods, ways to have fun, greater access to information, choices in ways to communicate, more places to visit and so on.[1] In a thoroughly commoditized economy, however, the increasing number of options translates into increasing numbers of things that money can buy and continuous improvements of consumer goods. At the same time noneconomic goods grow relatively underdeveloped. This gap in development is continually reinforced and expanded, because people naturally prefer the more developed options when making choices for how to allocate their time, attention, and financial resources. This self-reinforcing gap eventually forces noneconomic goods out of the sphere of actual choices that most people have, relegating them to luxury status available only to the wealthy, who can afford the leisure to enjoy them.

In this chapter we briefly look at agriculture, health care, environmental protection, transportation, electricity generation, and science and academia and see a consistent pattern of development where consumer products become highly developed while systems, relationships, and process-oriented options for growing food, providing power, moving people and goods, and protecting the Earth all suffer comparative underdevelopment. Rather than focusing on service-providing systems as a whole (health care, nutrition, transportation, environmental protection, etc.), in a commoditizing economy we concentrate our attention and resources instead on those elements of the system that can be most readily commoditized. As a result, every service delivery system becomes distorted. The products are overdeveloped and the systems are underdeveloped.

4.2   Unfair competition and comparisons

To understand how this works, consider as a hypothetical example two competing schools using contrasting approaches to teaching. Both approaches are based in sound pedagogy. School A's program involves its students heavily in a wide range of equipment, educational materials, and computer access. The plan is to package and distribute a standard set of tools and methods nationally and internationally. School B's teaching methods involve intense teacher and student interactions. The program is based on an evaluation of each school and a specific adaptation to the needs of each school, each teacher, and each student. At the heart of the program are community-building exercises and relationship development between and among the students and teachers. Once the system is fully tested, the teachers of B also hope to promote and spread the message of their successes with workshops and site visits.

Both schools see themselves as centers for research and development into their proposed teaching methods. They seek outside investments into their programs to allow them to carry out their R&D. School A receives significant investment from textbook, computer, and video distribution companies. Government officials, enamored of the idea of public/private partnerships, highlight the efforts of School A and match with government grants the investments from the private companies. School B, on the other hand, receives attention mainly from education activists, scholars, and teacher's organizations, and support from some community and private foundations. In the end, to test and refine their competing methods, School A receives annual investments of nearly $10,000 per student, while School B receives $1,000 per student. In the beginning, this is a roughly equivalent investment in teaching, because costs at school A are much higher, requiring considerable capital investment into materials and equipment. Over time, however, School B's costs grow much higher than A's because labor costs are higher. School A's is a capital-intensive method, School B's a labor-intensive one. In the beginning both schools show considerable promise. Both groups of students excel in a variety of standard tests. New investments pour into School A as companies are interested in spreading the word of this success. Likewise, teachers from School B make presentations at professional meetings and write articles for professional journals. However, there is little money for school site visits, and mounting labor costs are forcing cutbacks in both staff and materials.

School A is flourishing, School B is beginning to falter. Teachers' salaries rise at A, while at B, although a few dedicated leaders of the teaching method are in demand on the lecture circuit, teacher salaries are stagnant. School A continues to experiment with new and better equipment, while School B's experiments, which require hands-on interactions with teachers and students and intense observations, grow more infrequent. New developments in B are slow in coming. After a few more years, School A's methods have spread to many other schools around the world. Public/private partnerships are all the rage and schools adopting A's methods are high on the list of schools private businesses are willing to support. School B's methods are tried in several places, but in the age of reduced school budgets, labor-intense methods grow harder to justify. School A's methods are clearly successful, School B's show promise but flounder. Students at School A are scoring higher in standardized tests. Parents who can make choices send their children to School A. As a result, the school receives even greater financial support for its methods.

More schools adopt the methods. Most conclude that School A's teaching methods are superior. Few people bother with School B's methods any longer. Occasionally supporters of the B methods try to make a case for why their approach should be used. Small pockets of B teachers form networks of enthusiasts around the country. Occasionally their methods have a resurgence of support. There are newsletters and conferences and many small successes. Their adherents strongly believe that if everyone were just taught the benefits of B, they would rationally choose B over A. But over and over, A receives the overwhelming amount of R&D, its teachers make more money, and on and on. School B only has a chance if its proponents can find a way to counterbalance the overwhelming disadvantage they face in the competition for R&D investments.

It may well be that A is a superior pedagogy, but this can only be determined once the imbalance in R&D investment is corrected. Perhaps some synthesis of A and B may work best. But syntheses work best between methods that are at similar stages of development, or else the more developed will likely reassert its dominance over time until its partner is virtually eliminated. The original competition between A and B was measured by a criteria — average test scores or points per pupil — that was independent of the investment variable. Instead, if test scores were measured as points per dollar invested per pupil, the resulting evaluation would be far different.

This is precisely what happens in sector after sector in the competition between product-oriented and process-oriented methods, between capital-intense and labor-intense methods, between standardized and site-specific methods, between high and low commodity potential methods. High commodity potential satisfactions for human needs receive far more R&D investment than low commodity potential alternatives. Over time they become more "developed," more successful, and more attractive when compared to the less developed alternatives. Those who promote alternatives must face the argument that their approaches are not economically competitive with the fully developed industrialized model. This unfair competition will necessarily continue to thwart the full development of alternatives until some mechanisms are in place to direct research and development attention toward process-oriented alternatives.

4.3   Industrial or high-input agriculture vs. low-input agriculture

The evolution of agricultural practices, particularly in industrialized economies, is highly susceptible to distortion by commoditization. Alternative, indigenous and organic methods of farming rely heavily on local natural resources, community networks of cooperation, and personal skills, each of which is difficult to buy and sell. Organic farmers' approach to controlling pests, weeds, and diseases is one of prevention rather than treatment. As in health care, prevention is invariably more difficult to commoditize than treatment. According to the Organic Farming Research Foundation:

> Organic farmers' primary strategy in controlling pests and diseases is prevention. Organic farmers build healthy soils — fertilizing and building soil organic matter through the use of cover crops, compost, and biologically based soil amendments. This produces healthy plants which are better able to resist disease and insect predation. Organic farmers also rely on a diverse population of soil organisms, insects, birds, and other organisms to keep pest problems in check.[2]

Such farmers rely much more heavily on goods and practices that have very little commodity potential compared to commercial fertilizers, insecticides and equipment. Consider the differences between products and processes in the agricultural industry:

Agricultural system products (high commodity potential)
Proprietary hybrid and patented seeds
Insecticides, herbicides, and fungicides
Commercial fertilizers
Farm machinery
Fuel
Farm management books and magazines, etc.

Agricultural system processes and skills (low commodity potential)
Soil maintenance
Water conservation and management
Knowledge of soil, climate, local pests
Energy conservation and management
Nutrient cycling and enhancement
Crop rotation and placement
Rural networks of mutual aid
Pest control and management

Commoditization operates at the level of inputs and practices, preferentially developing the inputs that can be purchased, and tools and techniques that best utilize these inputs. Commoditization also operates on the outputs, favoring standardized crops marketed broadly, and selecting out varieties with characteristics that are adapted to unique soil and farming circumstances. The characteristics of the land itself are homogenized to accommodate the homogenized crops grown on it, so that the same crops can be grown in many places regardless of their original ecological characteristics. In this way the qualities associated with commodities are maximized: long shelf lives, portability, standardization of products, and industrial-scale production.

Alternative approaches to agriculture, sometimes lumped together under the heading of "sustainable agriculture," suffer from the same fallacy of unfair comparisons as the hypothetical educational approaches described above. The highly developed, commoditized agriculture sector brags of increasingly high productivity, to which the underdeveloped alternatives are then negatively compared. This is even worse when industrialized agriculture is compared with indigenous agricultural practices. These comparisons are made even more difficult and unfair as a result of the kinds of land each has been practiced on, particularly in countries with large inequalities of income distribution where the most fertile land is owned or controlled by the elite. Small and subsistence farmers are relegated to the least fertile, most inaccessible and steep uplands. These farmers are also the ones least likely to adopt modern farming methods and are more likely to continue the traditional practices they learned from the previous generation. So when comparisons are made between the productivity of traditional and modern agricultural sectors, what is compared is not only the results of productivity differences between farming practices but also the entire history of unequal access to land and resources. It is all the more remarkable that traditional farming practices have survived at all in the competition with highly commoditized industrial agriculture.

Little attention has been paid to alternative indicators of agricultural efficiency and productivity such as yield per volume of fuel expended, yield per inch of topsoil lost, yield per pound of pesticides applied. Such comparisons would favor those agricultural methods that are less energy intensive, protect and enhance soil health, and apply less pesticides. Each of these indicators would improve with decreasing use of commercial farm inputs and that, precisely, is their problem. The agribusiness industry is not concerned with developing ever-improving systems of farming, but instead exists primarily to market agricultural commodities. This can be clearly seen in the steady decline in the percentage of the food dollar that makes its way to the people who actually farm. In 1997 consumers spent $561 billion on food. Of that, 79% went for transporting, processing, and distributing the food. The remaining 21% was the gross return to the farmer.[3] Farm returns as a percentage of retail price has decreased every year from 1967 to 1997.[4]

As a result of the economic distortions brought about by commoditization, the farmer is at a distinct disadvantage compared to other participants in the food economy. What farmers control — land, soil, knowledge, labor, skills, and ties to land and community — have intrinsically low commodity potential. Farmers themselves, no matter how much they adopt the industrialized model of farming, are always at an economic disadvantage in a totally commoditized economy. Farmers who opt out of the industrial agriculture model are at an even greater disadvantage. The economic benefits of agriculture accrue most to those who control the key agricultural commodities: the crop once it leaves the farm and the key agricultural inputs, especially fuel, fertilizers, pesticides, and farm machinery. The economic interests that control the most commoditizable sectors of the food economy also control the direction of agricultural research and development. With stronger ownership rights in intellectual property by the seed and agricultural input industry, these private interests have become much larger players in agricultural research, which was once dominated by publicly funded agencies.[5] This leads to greater commoditization of the agricultural sector. This will continue unless and until society intervenes and insists that farming should achieve additional development objectives besides corporate profit and simple economic growth. Advocates for sustainable agriculture have developed checklists for assessing the sustainability of agricultural practices (see Table 4.1). A common feature of all the assessment criteria is that each involves skills, tools, and approaches with inherently low commodity potential. As a result, no matter how beneficial the practices are for the long-term sustainability of the world's food production systems, unless the pressures of commoditization are eased through deliberate public policy in support of sustainable agriculture, then farming practices will remain underdeveloped according to these criteria despite the best intentions and efforts of the advocates for sustainable agriculture.

Another way of looking at the commoditization of agriculture is as the systematic bias for products rather than processes. Consider alternative approaches to overcoming the major obstacles to successful crop production. These constraints include not enough of the right kind of nutrients at the right time and place, poor weather, insect pests, weeds and disease, too much or too little soil alkalinity, too little or too much water, not enough skilled labor to run the farm. There are two fundamentally different approaches to solving

Table 4.1 Checklist for Assessing Agricultural Practices for Sustainability

---

✔ Does the technology maintain/enhance soil quality?
  • soil lime
  • soil fertility (macro and micronutrients)
  • nutrient balance (macro and micronutrients)
  • soil structure
  • water-holding capacity
✔ Does it recycle nutrients?
✔ Does it prevent/reduce soil nutrient loss?
  • maintain soil cover
  • complementary root structure
  • water conservation
✔ Does it enhance crop diversity and protect diversity in surroundings?
✔ Does it enhance/maintain adequate perennial biomass (grasses, shrubs, trees, animals)?
✔ Does it use water efficiently and safely?
  • water use efficiency of crops
  • sustainable pumping rates
  • drainage that removes excess water while conserving soil
✔ Does it minimize or eliminate the use of toxic chemicals?
✔ Does it enhance the health benefits of food crops?
✔ Are maintenance costs (ecologically and economically) affordable?
✔ Does it recycle capital?
✔ Does it have neutral or positive effects on systems beyond the farm (watershed, village, downstream areas, nation, etc.)?
  • use of nonrenewable resources
  • pollution of air, water, soil and production of greenhouse gases

---

Source: Adapted from van der Werf, E., Can ecological agriculture meet the Indian farmer's needs? in *Let Farmers Judge*, Hiemstra, W., Reijntjes, C. and van der Werf, E. (Eds.), Intermediate Technology Publications, London, 1992, p. 207.

these problems: the product approach or the process approach. A third way, the approach advocated throughout this book for every economic sector, involves a balance and an optimization of both approaches. In the product approach, you purchase products that supply the nutrients your soil lacks, kill off the weeds and pests, adjust the soil pH, deliver the water, reduce the need for labor. With this approach, crops are bred to concentrate all their energy on the edible portion of the plant. The energy the plant would otherwise utilize for capturing nutrients, producing and distributing seeds, and other functions are supplied instead by purchased inputs. This phenomenon accounts for the dramatic increase in the amount of energy used to grow food. In contrast to the product-oriented approach, the process-oriented approach focuses much less on purchased inputs and much more on the natural processes of soil production and maintenance, recycling of nutrients, careful selection of locally proven seed stocks, and observation and understanding of the relationships between farmer and field. With the product approach the farmer engineers the environment to suit standardized farming methods while with the process approach the farmer engineers the farming methods to adapt to the environment. Both approaches require highly developed scientific observation and technical ingenuity. With the product approach this occurs mostly in the laboratory; with the process approach it occurs mostly on the farm. Neither approach is inherently more advanced than the other. With the product approach the focus of research and development attention is primarily on maximizing short-term economic returns on non-farm investment, whereas process research is focused on long-term sustainability of the farm. Used alone, the product approach is ideally suited for producing agricultural commodities for an anonymous market; the process approach is suited for producing food for a stable community. The product approach has led inexorably to a loss of agricultural diversity[6]; the process approach has led to an abundance of diverse crops. In combination, the product and process approaches are ideally suited to meet the complex requirements of contemporary food production for both farmers and consumers.

By all accounts, we want a balance of both highly commoditized agricultural inputs *and* less-commoditized agricultural skills and practices. But commoditization, by continually privileging goods with high commodity potential, invariably distorts agricultural development away from the ideal balance toward a highly industrialized, highly commoditized form of agriculture that has crowded out and even destroyed agriculture associated with the process approach. By initiating and adopting agricultural policies that deliberately promote research and development of the process approach and support alternative agriculture in its many forms, the pressures of commoditization can be balanced, and an agricultural system could emerge that reduces the dependence on purchased inputs, maintains the productivity and long-term health of the land, supports vibrant and prosperous farm communities, and establishes mutually beneficial ties between urban consumers and rural producers. For insight into what is possible with a process-oriented sustainable agricultural system we can look at some of the ways various indigenous peoples have managed their food production systems.

4.3.1 Indigenous agricultural systems

Agriculture most likely became necessary for survival once early humans had gotten so good at hunting and gathering that they began to deplete their immediate surroundings of readily available food. People learned about soil fertility and how it becomes depleted over time and how it can be restored. They gradually understood which crops required how much water and sunlight and how to optimize the delicate balance between the two. They learned to select the seeds from the tastiest, most nutritious, hardiest plants to sow again the next season and thus gradually bred crop plants with just the desired characteristics. They selected some plants that could be eaten immediately and some that could be stored and easily transported. With the beginning of trade, storage and transport became increasingly important.

As agriculture progressed, surpluses became large enough to feed people who did not need to work the fields, freeing human energy, attention, and creativity for art and industry. Agricultural productivity was one of the four primary routes to a prosperous, settled life, what we call civilization. The others were trade, conquest, and colonization. Interestingly, the characteristics associated with commoditization become increasingly important as you move from self-sufficiency to the voluntary interdependence of trade and then to the forced dependence of conquest.[7] Conquerors tend to concentrate on loot rather than yield. In many ways, as we shall see more clearly in Chapter 6, commoditization is the heir to the legacy of conquest and the systematic preference for loot over yield.

Throughout the world societies have developed through production and industry (agricultural and otherwise), trade, conquest, and colonization. Anywhere human development has occurred it has depended on those people who have worked the land and invented systems of agriculture uniquely suited for each of almost every type of soil, climate, slope, and pattern of rainfall on Earth. Almost everywhere on Earth, indigenous methods of growing food have been developed in a coevolutionary manner between local environments and local cultures over hundreds and sometimes thousands of years. Cultures have incorporated the knowledge of how to live in a particular place into ceremonial rituals and social norms, provided means to pass on information over long periods of time and allocated labor and distributed the harvest.[8]

None of these indigenous agricultural societies were idyllic — far from it — but many evolved locally appropriate practices and tools whose evolutionary development slowed and sometimes abruptly ended with conquest, colonization, or overly rapid environmental change (sometimes precipitated by population growth resulting from agricultural successes, sometimes brought on by the environmental ignorance of colonists or conquerors). It is impossible to know what might have happened if indigenous horticultural practices had been able to continue to mature and develop. Most were wiped out by conquest, imperialism, and colonization when the new settlers planted their own agriculture or transformed subsistence agriculture into commodity-based export agriculture. Who knows what might have developed in the Americas, for example, if the native agronomy had continued to develop and native farmers had been able to continue experimenting and innovating, or if contact with European settlers had been a true sharing and synthesis of knowledge and skills.

What did happen shows the connection between commoditization and conquest. The conquering European powers appropriated the *products* of indigenous agriculture, which then became the basis of modern industrialized agriculture, but ignored or even suppressed the *processes*, the locally specific tools and techniques and the cultural inclination toward communal responsibility and conservation that had created the conditions for the horticultural innovations in the first place.

The least commoditizable aspects of native agriculture were those aspects most closely associated with the cultural and spiritual life of the community. American Indian culture underwent catastrophic changes as the natural wealth of these communities was expropriated by Europeans, and the *products* of indigenous agriculture were adopted and spread worldwide, greatly advancing the material development of the European powers. These new crops were able to flourish in difficult soils, which allowed European farmers to plant previously unproductive lands and resulted in a nutritional bonanza of unprecedented proportions. The better-fed European population experienced a jump in life expectancy and rapid population growth. In addition to the calorie-rich supercrops of corn and potatoes, seventeenth and eighteenth century Europeans also adopted a multitude of other Native American crops, including sweet potatoes, tomatoes, pumpkins, gourds, squashes, watermelons, beans, certain grapes and berries, pecans, black walnuts, peanuts, maple sugar, tobacco, and cotton. Certainly in terms of variety of crops and their caloric content, the exchange between Europe and the Americas benefited the Europeans far more than the indigenous Americans. By this reckoning, agriculture and by presumption many other aspects of culture were far more advanced in the New than the Old World.

A look at the achievements of indigenous farmers around the world gives a glimpse of the agricultural diversity and productivity that is possible when development is not driven exclusively by commoditization. Three thousand years ago Incan farmers were able to grow up to ten tons of cereal grains per hectare at high elevations where modern techniques using chemical fertilizers have been able to produce no more than four tons per hectare. The Incas planted crops in long and narrow raised beds, which they constructed with the nutrient-rich sediments they regularly retrieved out of their extensive system of canals. The Spanish invaders reported that the natives grew food for the townsfolk in "floating gardens" in lakes throughout the city. Little is known about the techniques or the products of these gardens, but they suggest a highly developed agricultural system extensively integrated with urban design.

Throughout the Americas different crops were bred for different growing conditions. One plant could survive the floods, another the drought. One was susceptible to beetles, another to ants. Diversity is crop insurance of the most fundamental kind. The diversity of crops produced by the some of the indigenous farmers of the Americas was the outcome of long periods of experimentation, observation, and cultivation of plants. The Spanish conquistadors recorded their amazement and disgust (because of the unusual things the natives ate) at the diversity of the Incan diet. Incan farmers had cultivated around 70 different edible plants and gathered more than 600 species of wild plants, 300 species of fish, and over 100 kinds of insects.[9]

The food production and preparation methods of traditional indigenous peoples are often more productive in terms of calories and protein per unit of area, but less valuable in terms of yield of commodities per unit of area. For example, studies have compared the farming practices of the Kayapo people of the Amazon with those practices brought into the rain forest by colonial farmers and ranchers who more recently cleared and settled the forest. The Kayapo system relies on a diverse selection of crops that provide both food and raw materials for their way of life, and maintain soil fertility for long periods of time. Their land management practices produce more resources and protein at lower environmental cost than either the colonial crop or livestock production systems. Edible harvest and protein yields were measured and compared over 5 and 10 years. The Kayapo yields were reported to be nearly triple those of the colonial farmers and nearly that much greater than those of the ranchers. As it turned out, the researchers were unable to compare the production of Kayapo agriculture to the small farms beyond the 5-year period, because most colonists had exhausted their land within the 5-year period and the farmland had been abandoned. Although cattle grazing is a somewhat more sustainable use of cleared Amazon land than is colonial crop production, the comparison between the Kayapo and livestock systems shows that the ranchers were able to produce a mere 700 kg of stock per hectare, whereas the Kayapo system produced 84,000 kg of mixed crop. Even when comparing just protein yield per hectare, the Kayapo growers produced from vegetable sources alone roughly eight times per hectare what the protein ranchers got from their meat.[10]

Despite the comparative yield advantages of the Kayapo system, it cannot compete economically with the colonists. Livestock can produce greater profits in a much shorter period of time. As is usually the case, a high rate of return comes at the cost of the land. The ecological devastation of transforming rain forest to pasture has been well documented.[11] After the Amazon is largely destroyed, the investors who have turned a profit will move on to other opportunities. The Kayapo will not be able to move; nor will they be able to maintain their traditional way of life.

Without the pressure to make land maximize income, perhaps a balanced and reasonable approach to farming the Amazon could be developed. Such an approach would begin with the Kayapo's knowledge and skills about producing food and resources in the Amazon to build and sustain jungle communities. In order for that to occur, some kind of countervailing pressure to commoditization would have to be applied. The main difference between the colonists and the Kayapo land management systems is that the former is organized to produce a product in the cash economy, the latter is meant to produce food and resources to sustain a way of life. The Amazon is so rich in resources that both commodities for the world market and sustenance for the indigenous peoples could be produced from it.

When the National Research Council undertook a study of agricultural systems in the world's rainforests, it concluded that

> [A]gricultural systems and techniques that have evolved from ancient times to meet the special environmental conditions of the humid tropics include the paddy rice of South-East Asia, terrace, mound and

> drained field systems, raised bed systems (such as the
> chinampas of Mexico and Central America), and a va-
> riety of agro-forestry, shifting cultivation, home gar-
> den, and natural forest systems. Although diverse in
> their adaptations, these systems share common ele-
> ments such as high retention of essential nutrients,
> maintenance of vegetative cover, high diversity of
> crops and crop varieties, complex spatial and temporal
> cropping patterns, and the integration of domestic and
> wild animals into the system.[12]

These common elements derive from site-specific knowledge, experimentation, observation, and integrated social systems that have evolved in conjunction with unique local ecosystems. Although there are countless simi-
larities between human cultures everywhere, each has evolved in response
to unique challenges posed by a particular environment. The NRC study
concluded with concern that the diversity of cultures is being lost in the
tropics as fast as, and often faster than, biological diversity. Traditional
knowledge, like biodiversity, is not a product that can be sold, and therefore
will lose out in the economic competition for survival as long as self-regu-
lated markets are expected to be the primary means of allocating research
and development attention and economic resources.

   High levels of productivity and diversity are only obtainable when people have intimate knowledge of the plants, animals, climate, and growing
conditions of a particular place gained over a long period of observation
and experience. This knowledge is often encoded in and reproduced through
the cultural practices of the people. It is their most valuable capital. This
means that for indigenous peoples the loss of their land results in the loss
of the culture itself. Unfortunately, the rate of cultural annihilation is increasing with the rate of commoditization around the globe. This is a tragedy not
only for the peoples whose cultural life is wiped out, but also for all of us
who must suffer the loss of human knowledge, capacity, and potential.

4.3.2   *Productivity and unfair comparisons*

According to the conventional wisdom, even though alternative and indigenous forms of agriculture may be better at reducing exogenous inputs and
preventing soil erosion, industrial agriculture produces higher yields.
According to this reckoning, the cost of a large-scale transition to sustainable
methods would be simply too costly and dangerous to be worthwhile. This
assumption is misleading.

   Industrialized agriculture is indeed more productive in growing the
commoditized economy, but not necessarily in growing food. There are
indeed economies of scale that can be achieved by industrialized farms, but
these savings are realized by improved labor productivity, not necessarily
land productivity. In terms of land productivity, small farms, which rely

heavily on skilled labor, can yield more crop per acre than large farms. The most productive farming turns out to be small, labor-intense, garden-like cultivation systems with mixed crops, shifting cultivation, and a high degree of nutrient recycling. Such systems are capable of producing three times as much crop per unit area as highly mechanized, capital- and energy-intense agricultural production systems utilizing minimal human labor.[13] Industrial farms are far more productive in terms of yield per worker–hour, whereas low-input agriculture is more productive in terms of yield per area, and far more productive in terms of yield per unit of any of the major inputs such as chemical fertilizer, pesticides, fuel, and irrigation water.

For farms larger than intensively cultivated gardens, small farmers produce about twice as much per hectare as do large, industrial-scale farmers, while using only one fourth or one fifth as many purchased inputs.[14] Studies in modern Mexico have shown the advantages of traditional companion planting methods. These methods rely on the mutually beneficial characteristics of corn, beans, and squash. The corn stalk provides structure on which the bean plant can climb, while the broad squash leaves shade out weed growth. Insect-attracting plants are deliberately planted at the edge of the fields to draw pests away from the crop, while other plants are cultivated in the field to deliberately attract those insects that feed on the most damaging of crop insect pests. Fields planted in this diversified pattern have been able to increase yields by as much as 50% over the conventional method of planting a single, highly marketable commercial variety.[15]

Most traditional farming methods rely heavily on skilled labor. But labor in a commoditized economy is increasingly expensive relative to those inputs with higher commodity potential: machinery and raw materials. Since some of the highest costs of any operation are associated with labor, particularly skilled labor, investments naturally focus on reducing costs through mechanization and standardization. By eliminating skilled labor, you eliminate the very resource most essential to sustainable agriculture — people with intimate, detailed knowledge of particular lands and soils. A single technician giving instructions to untrained field hands on an industrial farm can manage several thousand acres using standard recipes for determining the amounts and timing for the application of pesticides and fertilizers. Organic and other forms of sustainable agriculture, on the other hand, require highly trained and experienced labor and diverse management skills. Because resources for agricultural training are distorted by commoditization, there are almost no formal training options available for individuals interested in obtaining the skills for sustainable farming. Internships are few, college training programs even fewer.

One example of how this lack of skills and training affects agricultural practice has been the disappointing progress in implementing Integrated Pest Management (IPM) programs as an alternative method of pest control. In the 50 years since farmers started using DDT for protecting crops, use of pesticides has grown 33-fold. Worldwide, 50 million kg of pesticides were applied to crops in 1945; by the mid-1990s annual worldwide pesticide use had jumped to 2.5 billion kg.[16] Pests continue to develop resistance and become even more harmful as a result, leading to increasing pesticide usage and the continuing need for new and more targeted pest management products. All the while, pesticides continue to spread into the water and air, endangering other plants and animals as well as humans. The problems with increasing reliance on chemical pesticides has been known for decades. Excellent alternatives have been developed, most under the heading of IPM. These systems are designed to support and promote natural pest resistance through companion planting, hardier breeds, introduction of beneficial insects and pathogens, and the controlled use of narrowly targeted and nonpersistent pesticides. Despite some considerable success with these methods, they have not been widely adopted, except on a few crops, and have not significantly affected the amount of pesticides used worldwide. The failure to adopt IPM stems from two facts: the full cost of pesticides in terms of environmental health effects are not reflected in their price, and the implementation of IPM requires careful management, experimentation, and observation — all noneconomic goods and services with low commodity potential.[17] IPM, like other efforts designed to decrease the environmental impacts of farming, require just the kind of site-specific knowledge and understanding that is in decline worldwide.[18]

Commoditization consistently distorts development toward capital-intense and less productive large, industrialized farms that are devoted to the production of a single or just a few crops. Cropland devoted to monocultures has expanded worldwide, where a single large crop (usually wheat, corn, or rice) is grown year after year with heavy inputs of pesticides and fertilizers. At most, some land is simply rotated between corn and soybeans. Farm animals are increasingly raised apart from croplands, further reducing the opportunities to recycle animal waste as soil nutrients and greatly adding to the pollution load on streams and rivers near chicken houses and feedlots. The number of hog and dairy operations declined by 70% from 1969 to 1992, while production, now highly concentrated, remained stable.[19] At present, 90% of the world's food supply comes from only 15 species of crop plants and 8 species of livestock among the estimated 10 million species of plants and animals in the world.[20]

The way agriculture is commoditized is through the replacement of human labor, animal power, and renewable energy with fossil fuel energy. Each calorie of food we eat from high-input agriculture embodies several calories of fossil fuel energy. From petroleum comes the synthetic chemicals in pesticides. Oil powers the production of chemical fertilizers, moves agricultural commodities around the world, drives the farm machinery, raises the irrigation water, and on and on. High-input agriculture requires about 3 kcal of energy derived from fossil fuels for every 1 kcal of human food produced. Food production accounts for about a third of all energy used. The U.S., the most commoditized of modern economies, uses three times as much energy per capita for food production than developing countries use for all energy-consuming activities combined, including food production.[21]

This measure, the amount of food energy produced per unit of energy expended in food production, is a measure of energy efficiency. The energy efficiency of modern conventional agriculture declined in the U.S. throughout the period 1920–1973.[22] High commodity agriculture is tremendously inefficient in terms of energy. Consider rice production. Modern rice farmers get a negative 1 to 10 energy return. In other words, they use up to ten times as much energy to produce the food as the resulting food yields in calories.[23] Compare this, for example, to the traditional rice farmers of Bali, who are reported to produce 15 calories of food energy for every 1 calorie of energy used — and even higher yields are sometimes obtained.

Almost all the negative environmental consequences of conventional agriculture result from this massive dependence on nonrenewable fossil fuel energy. As we have seen, most of this energy use does not go into increasing agricultural production as much as it goes into agricultural commoditization. The key to reducing energy dependence in agriculture lies in providing a balance for commoditization. Much of this balance would take the form of supporting the development of sustainable alternatives to industrialized and highly commoditized agriculture.

Because of commoditization, organic agriculture in general cannot compete for private investment capital. Up-front public funding of research, development, and agricultural extension is essential for organic farming to succeed. To achieve its full potential organic farming requires time and the knowledge and skills of the farmers who build up the humus content of the soil. Without public financial support it is extremely difficult for a farmer to stick out the lag time between start-up or changeover from standard farming practices to organic practices. In addition, organic farming is very labor-intensive, and the costs of sufficient labor can be prohibitive.[24]

When farmers switch to organic methods, they experience a short-term decline in productivity for some crops in terms of yield per unit area and a dramatic rise in productivity in terms of yield per unit of energy input. A 1980 USDA study, however, not only found increased energy efficiency on organic farms but roughly equivalent yields per acre of organic vs. standard methods. For wheat there was no significant difference, for soybeans, organic methods produced 14% higher yields. Most studies show a 20 to 30% decline in productivity in the short term, with productivity increasing slowly but steadily in organic farms over time.[25]

Agricultural research dollars go overwhelmingly to benefit the dominant agricultural paradigm. The Organic Farming Research Foundation analyzed the U.S. Department of Agriculture (USDA) Current Research Information System to assess the pertinence of the research to organic farming.[26] Of the 4,500 projects the authors reviewed, only 301 led to results that were pertinent to organic practice. Funding for these projects amounted to about 0.1% of the USDA's annual research and education budget. In another study, the CRIS database was searched for projects that included the words "sustainable" or "low-input" in their title or abstract. These 122 projects were then analyzed for content. According to Molly Anderson, who cites the studies, "Few of the

122 projects showed the broad scope that writing about alternative agricultural research emphasizes. Only 22% dealt with entire farms, 25% looked at
both crops and livestock and 19% studied general processes from which basic
agroecological principles could be learned."[27] Both these studies analyzed the
CRIS database, which includes only publicly funded research and probably
greatly *over*estimates organic's actual share of total agricultural research. Privately funded research tends to be even less relevant to organic farmers and
other forms of alternative agriculture. Figures 4.1 and 4.2 compare the amount
of funding directed toward research relevant to organic agriculture with total
USDA research funding and total private funding.

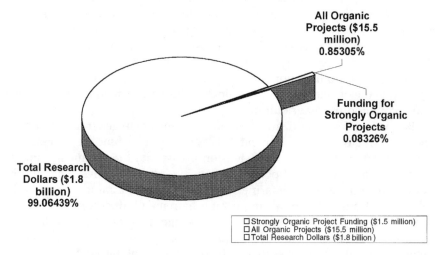

Figure 4.1 Comparison of dollars spent on "strongly organic" projects vs. all other
projects by USDA in 1995.

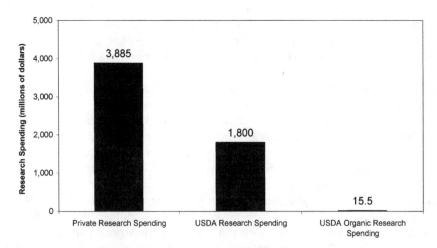

Figure 4.2 Agricultural research expenditures by private and government sectors
compared to investments in organic farming research by USDA in 1995.

Instead of comparing averages, the performance of the *best*, most productive low-input farmers should be compared with the best, most productive high-input farmers. In general the best, most experienced organic farmers can achieve yields equal to or better than conventional farmers while decreasing energy use, building up humus in the soil, dramatically reducing soil erosion, reducing nitrate concentrations in groundwater, and enhancing plant resistance to disease.[28] Studies undertaken by the Rodale Institute concluded that experimental plots using standard organic farming techniques had yields similar to plots using conventional methods over a 15-year period. While the experimental crops produced as well as the conventional plots, they also enriched the soil. Carbon levels rose dramatically and nitrogen losses were half of what they were in the conventional high-input plots.[29] Well-maintained soil acts as a carbon sink, preventing large amounts of carbon from entering the atmosphere, not inconsequential when industrialized societies are straining to figure out ways to reduce the increase in carbon dioxide in the atmosphere in order to stem global warming.

The stakes continue to grow. It is going to be increasingly difficult to raise yields enough to meet the needs of growing populations when the amount of cropland is declining on a per capita basis. Farmland is disappearing under human settlement and environmental degradation. Although new areas continue to be opened, the most fertile lands were long ago put under the plow, and much was later then buried under cities and suburbs. Since the 1950s, total land area in grain cultivation has increased by around 19% while global population has increased 132% resulting in a decline of 50% in the amount of grain area per person.[30] Heavy uses of fertilizers and pesticides have more than doubled average yields, so that per capita production has continued to increase. We have begun to see potential saturation points beyond which increasing fertilizer use will not likely increase yields to the same degree as initial applications once did.[31]

Calls for the continued intensification of agriculture are likely to increase along with charges that proponents of alternative, low-input agriculture are essentially irresponsible. The results of the "unfair comparison" are only likely to worsen. We approach what looks like a choice between feeding growing populations of hungry people and protecting the environment. This is a false choice.

All agriculture disrupts natural ecosystems. That is what agriculture is meant to do. However, disruptions can be limited, first of all by limiting the number and amount of agricultural inputs such as fertilizers and pesticides. While it may not be possible to maximize both environmental protection *and* crop yield, each can be optimized in relation to the other. The goal of agricultural policy and practice should be to optimize harvest *and* ecological integrity, which will maximize neither. This will work only within an economic context that balances commercial and noncommercial values. Economic reform (as described in Chapter 8) will be required.

Even without systematic economic reform, consumer demand for organic food will likely grow, and the market niche for organic products will expand. This is consistent with the commoditization of alternative agriculture. Commoditization works at all levels of commodity potential. So-called alternative or organic agriculture is equally subject to the process of commoditization. One approach to developing the low input sector is to serve "specialty markets" of consumers willing to pay higher prices for organic agriculture, as well as specialty services for the growing market for organic gardening supplies. As a result, the investment capital that is available for low-input agriculture goes preferentially to the production of commodities for the organic agriculture specialty market. Investments of this nature are absolutely essential if organic farming is going to survive. But they will do little to move the greater portion of agriculture in a more ecologically friendly direction. A real and widespread change in agricultural practices will occur only if economic incentives and R&D investments are deliberately redirected toward processes such as nutrient recycling, integrated pest management, water and other input conservation, and local site-based organic experimentation rather than particular end-products.

In a poll published in February 1997, conducted on behalf of the Swiss-based biotech and pharmaceutical giant Novartis, 54% of American consumers said they would like to see organic food production become the dominant form of agriculture in the U.S.[32] The preference of the majority of citizens is not for a small selection of higher priced foods, but for a transformation of agricultural methods. The potential market is there, the skills and techniques are being developed; what stands in the way of organic agriculture becoming the dominant form is commoditization, which preferentially develops the methods of factory farming. This makes the large-scale transformation of agriculture "impractical" and "uneconomic" under current circumstances.

Yet in the U.S. and other wealthy nations, the market for organic products has grown large. The U.S. organic foods industry recorded $3.5 billion in sales in 1996, with average growth in sales of 20% annually since 1990.[33] The existence of a growing market for organic foods represents a powerful critique of industrial agriculture that includes concerns about:

1. The presence on food of the residues and by-products of agricultural inputs, especially pesticides on grains and vegetables and growth hormones and antibiotics in meat and milk;
2. The poor treatment of animals confined in crowded factory farms;
3. The presence of food additives and irradiation byproducts;
4. The industrialization of farming and the continuing decline in small farms and farming communities;
5. The environmental impact of large, concentrated feedlots and other factory farming practices;
6. The extension of cultivation and ranching into forests and wetlands;
7. The integration of food production, processing, and marketing in global industry with the subsequent demise of regional and local markets and symbiotic relationships between urban areas and agricultural surroundings;

8. The increasing development and use of crops whose genes have been altered by genetic engineering;
9. The dangerous narrowing of the diversity of food crops;
10. With globalization of markets, the increasing difficulty of tracing sources of products and monitoring food safety.

However much the public resents the environmental and health costs of industrial agriculture, changing to organic agriculture will be difficult. It will necessitate public investment, and government support for organic agriculture, not merely to support the growth of an emerging specialty market, but to actually support and nurture the less commodity-intensive goods and services upon which organic agriculture can flourish: knowledge of and experimentation in local environments, improvements in nutrient recycling, planting, and harvesting methods (not just tools), local farmers' markets, and a massive agricultural extension program and consumer education infrastructure dedicated to ecologically sound farming practice.

The growing public awareness of these issues, especially food safety concerns, creates a growing consumer demand for alternatives. This demand poses a fascinating dilemma for the commoditized economy: how to satisfy this market demand without actually giving credence to the concerns about and critique of industrialized agriculture that lie beneath the consumer demand. This dilemma exploded into political controversy in the winter of 1997/98 when the U.S. Department of Agriculture issued its proposals for developing a national set of standards for use in certifying practices and products as "organic."[34] Many food activists believed that adoption of the USDA proposals would allow a situation in which products would be labeled "organic" even though they were produced through practices that most activists would disallow. These included the nonmedical use of antibiotics as animal growth stimulants, genetically engineered products, the crowded confinement of farm animals, use of nonorganic animal feed, and other practices associated with factory farms. Over 200,000 public comments were generated by food activists, 90% opposed to allowing these practices. Perhaps most troubling of all for opponents, the proposed Certified USDA Organic label would, if adopted and implemented, reserve for the USDA a legal monopoly on the word "organic." Independent certification programs that demanded stricter standards to qualify for labeling could be subject to civil suit. As of this writing, the USDA has temporarily withdrawn its proposal and is considering revisions.

The purpose of certification and labeling is to standardize the meaning of "organic," to minimize confusion among consumers and protect organic farmers from the deceptive tactics of those making unfounded and unchallenged claims about their products. There are two approaches to an organic certification program, which divide roughly along lines of high and low commodity potential and the distinction between product and process. One approach is to certify particular farms and growers based on the way they grow their crops. Rules cover soil protection and enhancement methods, crop rotations, pest control, density of animal confinement, fertilizer solubility. Such a certification program requires frequent auditing of practice through on-site observation. The commoditized version of organic certification concentrates not on monitoring farming practices, but instead on a technology-intense analysis for chemical-specific residues and other properties of final products. The latter version of "organic" not only fails to shift agriculture from high-input to low, but actually adds a whole other level of processing and energy inputs in the form of testing for residue-free products. A product-based certification program eliminates the need to concern oneself with farm practices and overlooks genetic engineered alterations, irradiation in food processing, and the use of sewage sludge for fertilizer — practices that are rejected by most of the 28 private and state-level programs that presently certify organic farming practices.

Consider two facts: first, as noted earlier the most productive agriculture turns out to be small, garden-like plots intensively managed by small landholders. Second, the largest single crop in the U.S. turns out to be lawn grass, which now covers 25 million acres. People caring for their lawns apply three times more fertilizers than do farmers on their crops.[35] What might be possible toward community food self-sufficiency and environmental improvement is if a substantial portion of the lawn acreage was devoted to food crops. Experiences during war times have shown that individuals readily take to the challenge of gardening. What is required is neighborhood coordination; public, shared food processing centers; decentralized agricultural extension; and community building — all of which rank low in commodity potential.

4.3.3   Research and development

Investment in research and development probably plays the greatest role in agricultural commoditization, indeed all commoditization, by disproportionately investing in R&D that yields new and better commodities, not necessarily new and better agriculture. Take, for example, the way in which genetic engineering technology has thus far developed.[36] Plant modification through manipulation of genetic material theoretically has the potential to improve agriculture in many ways. It is possible to conceive of plants engineered to resist or tolerate conditions that otherwise would cause damage: drought, nitrogen limitation, saltiness, cold and heat, toxicity. Genetic engineering holds the promise of rapidly doing what traditional agronomists did over centuries: developing a diverse range of crops uniquely suited to local conditions. R&D investments are largely ignoring this potential and have disproportionately been invested (40% by 1996) in herbicide-resistant plants. By engineering this resistance, farmers will be allowed to use more herbicides without damaging the crop. This helps sell more herbicides, while increasing chances of developing herbicide-resistant strains of weeds, triggering even greater increases in herbicide usage.[37] The companies that control this investment also largely control the herbicide market and from their perspective, this is the best use of investment dollars.

More recently, two other highly commoditized areas of research have dominated R&D in genetic engineering. One is the introduction of a gene from a bacterium that produces a natural pesticide (Bt). This gene has successfully been introduced into cotton plants, which in effect now produce their own pesticides. Another area of development has been in the so-called Terminator gene, a genetic manipulation that makes the seeds produced by crop plants sterile, thus making it impossible for farmers to collect seeds after harvesting crops. These three technologies, along with the genetically engineered growth hormone that increases milk production in dairy cows, have dominated the early agricultural product development in applied genetic engineering, and each greatly increases the pressure of commoditization and the control that private corporations have over the world's farmers.

The world is in a bind. World per capita food production has increased by 15% from the 1960s to the 1990s. But per capita figures are misleading. Modern commoditized agriculture has been enormously successful in producing a wide variety of food for the tables of those who can afford it. For those who are left out of that market for one reason or another, the result has been a devastating loss of capacity to achieve self-sufficiency in food. Current agricultural practices contribute significantly to all the major environmental problems facing the world: global climate change, loss of biological diversity, polluted and overdrawn water resources, spread of toxic chemicals, and air pollution. The combination of trends necessitates a shift in agriculture toward more benign and sustainable practices. In addition, agriculture in general suffers from problems of diminishing returns. The first crop planted in a cleared field yields more than subsequent crops. The initial application of fertilizers can boost production substantially, but later applications are less effective. The same is true for irrigation water. The first applications of pesticides are more effective than later ones, after pests begin to adapt and evolve pesticide-specific resistance.

The good news is that because of the underdevelopment of organic methods, we are a long way from reaching the limits of the potential gains to be had there. We have considerable knowledge of how to farm sustainably and an ancient history with many successes. The bad news is that commoditization pressures continue to skew investment toward research and development in the opposite direction of sustainability. In addition, ecological disruption changes the ecosystems within which traditional farming methods have evolved. Once these changes occur, traditional practices based on knowledge of local ecosystems are no longer viable. We are losing our inheritance. Global climate change makes this phenomena worse still. All agriculture depends on relatively stable climate; a shift into a new, more unstable climate regime would severely complicate our task. This is a phenomenon we will see in each sector that we analyze. Commoditization is preventing us from achieving sustainable development, and in the process grossly limiting development of our full human potential. The more extensive the commoditization, the more ecologically disruptive it is. The longer it lasts, the more underdeveloped alternatives become and the more difficult the road to sustainability becomes.

4.4   Health care and health services

The most commoditizable items in the health sector are pharmaceutical drugs. Not surprisingly, their sales in recent years have skyrocketed, along with the expenses of promoting them. In the U.S., between 1990 and 1998 the sale of prescription drugs more than doubled. In just 4 years, 1990–1994, the top 40 drug companies increased their salesforce by 60%.[38] There is a persistent belief in many parts of the world that access to health care should be a right, not a privilege reserved for those who can pay. As a consequence, there is always some effective resistance to its complete privatization and commoditization. Health care has been treated at least in part as a public good and a public service whose provision is the collective responsibility of society. This does not mean that health care is not distorted by the same sort of pressures that affect agriculture. In order to understand the effects of commoditization in health care, as in agriculture, a useful distinction can be made between products and processes. Health care *products* inherently have greater commodity potential than health care *processes*. Some examples of this distinction follow:

Health care products (high commodity potential)
Pharmaceutical drugs
Hospital/health care supplies and equipment
Vaccines
Diagnostic equipment
Vitamins and herbs
Health books and magazines
Exercise equipment

Health care system processes and skills (low commodity potential)
Disease prevention
Health maintenance education
Hands-on therapy and care
Doctor/patient interaction
Exercise/outdoor play
Personal hygiene/sanitation
Counseling

In the industrialized world, health care products abound; the processes of health enhancement and disease prevention are the poor cousins. For example, 99% of health expenditures in the U.S. went for medical treatment of individual consumers, while 1% went for public health initiatives directed to populations. Estimates made of how most early deaths can be prevented suggest that of those that are preventable 70% would best be addressed through public health and accident and disease prevention initiatives, while 10% are amenable to prevention by medical treatment.[39]

The R&D investment gap between products and processes in health care may be less than in other areas because of the political popularity of basic research into common devastating illnesses. But even in publicly financed research, the emphasis is overwhelmingly slanted toward elucidation of fundamental physiological mechanisms that lend themselves to manipulation by pharmaceutical intervention. Far less research support goes toward understanding the social, cultural, behavioral, and political aspects of health maintenance and disease prevention, even those that are known to play the most significant roles in health. In 1992, according to a paper by Dr. John Farquhar, $28.7 billion were spent on health research in the U.S. and about 9% of it was allocated to prevention research. This figure represents approximately 0.0032 of the total U.S. health care budget.[40] In contrast, in the U.S., the drug industry increased its R&D expenditures for new medicines from $6.5 billion in 1988 to an estimated $20.6 billion in 1998, approaching the total amount of all other health research combined.[41]

A good example of how commoditization pressures distort health care can be seen in the world's reaction to the HIV/AIDS pandemic. The United Nations Program on HIV/AIDS and many other health organizations have called for the adoption of three main strategies to reduce transmission of the virus:

1. Communication of HIV/AIDS information with the purpose of changing high-risk behavior;
2. Promotion and distribution of condoms and education on their use;
3. Better diagnosis, treatment, and prevention of sexually transmitted diseases.

Each of these approaches have low commodity potential, with the possible exception of condom campaigns. But the most effective condom campaigns involve their free distribution, and there is little room for the kind of value-added product differentiation that characterizes most commodities. Where the recommended HIV prevention strategies have been implemented with well-designed programs, they have proven successful in substantially reducing the number of new HIV infections. But despite these successes and the pessimistic outlook for an affordable cure or vaccine, it is not surprising to see that prevention strategies receive only a small percentage of HIV/AIDS funding worldwide and an even lower percentage of the research and development dollars allocated to studying AIDS.[42] The percentage of the AIDS budget in the U.S. devoted to prevention has been declining. It went from 19% in 1993 to 14% in 1998.[43]

There is a persuasive economic argument for prevention: it is an oft-repeated truism that an ounce of prevention is worth a pound of cure. Poor health hurts businesses through decreased worker productivity and high costs of employer-provided health insurance. Governments face huge costs as they attempt to subsidize or directly provide health care services. In the U.S. where health care services are the most commoditized, costs have been rising at an astonishing rate, quadrupling between 1980 to 1995.[44] As a result, powerful economic interests in favor of health maintenance and disease prevention should exist. But because of the difficulty in commoditizing prevention, what has occurred instead is the commoditization of cost controls through the innovation of the health maintenance organization (HMO). Between 1980 and 1995 the number of people enrolled in HMOs as their health care insurer rose from 9.1 million to 46.2 million.[45] HMOs, by profiting through cost-reductions, create a competitive group of private insurance interests that provide a countervailing force against the ever-increasing pressure to purchase the full array of health care products with the most commodity potential: drugs and equipment. The problem with the HMO approach to health maintenance is that it fails to redirect health care savings toward those preventative and health maintenance *processes* with low commodity potential: community health education, pollution reduction, increased prenatal care, improvements in sanitation, public hygiene and nutrition. Instead much of the savings are captured in the form of profits for the insurance industry and kept in the private sector where they are further invested toward commoditized economic growth in health care or the larger economy.

The great expansion of the insurance industry in modern economies itself is a fascinating case of the commoditization of mutual aid. Noncommoditized or "primitive" insurance consists of socially mandated norms that require community members to help out in times of illness or death in the family. Private insurance takes a set of mutual understandings, expectations, and relationships and transforms them into a product for sale. Insurance company receipts, since they do not need to be reinvested in the production of more of the goods the customer needs, can thus be invested in other profitable activities. This creates a huge storehouse of investment capital, which becomes part of the capital that disproportionately flows toward the production of commodities.

Hands-on or labor-intensive prevention and treatment has inherently low commodity potential and as a result will get short shrift in the commoditized health care industry. Diagnostic equipment to support a physician's observational skills (*products*) are highly developed, while the observational skills themselves (*processes*) grow increasingly and comparatively underdeveloped. This is also true of healing arts which, like agricultural practices, coevolve within a set of local dynamics peculiar to a given culture. Traditional herbal medicines, which relied on the cultivation and collection of local plants, receive far less attention from investors than do pharmaceutical medicines that can be patented. This has changed somewhat in recent years as a result of consumer demand for access to alternative health care products. Within the alternative medicine field, the standardization, packaging, and marketing of herbs has now become a commodity-intensive industry.

In the mental health field, far more is invested in research and development of psychoactive drugs than in effective forms of counseling or in mental health approaches involving peer support networks, exercise, or other life-style changes. In the health care industry, as in all businesses, productivity is measured as a unit of output per unit of input. There are also general measures of productivity such as incremental increases in average life expectancy per unit invested. Studies by American epidemiologist Charles Stewart compared the impact on life expectancy of various factors associated with economic development. What he found was that two factors — clean public water supply and literacy — accounted for 85.5% of the measured differences in life expectancy around the world.[46] Analyses of the United Nations human development report conclude that the two most important variables determining the quality of health in a country are the number of nurses (not doctors) per capita and the percentage of women who receive a secondary education. Similarly, despite several analyses that demonstrate that the impact of human population growth creates the most significant negative pressures on environmental and public health,[47] and that the education and social status of women is the leading predictor of fertility rates,[48] by far the largest share of investments in birth control has gone into fertility reduction products rather than fertility-reducing social change that benefits women. Most of the evidence suggests that the key to population stabilization — reduced fertility rates — is in near-universal literacy, equal rights and legal standing for women, and a social security net for the poor.[49] However, as with most commoditization effects, the emphasis has been not on system change but on pharmaceutical birth control products as the primary solution to excessive human population growth.

Highly commoditized health care runs into problems of diminishing returns on investment. Beyond a certain point, each dollar spent on health care yields less in terms of additional years added to life expectancy. The ratio of added life years to national health expenditures (as measured as percentage of GNP) decreased by nearly 60% between 1930 and 1982.[50] There is little relation between health expenditures and outcome, at least in terms of life expectancy.[51] Figure 4.3 shows the relationship between health expenditures and life expectancy in several countries. The U.S. spends nearly twice as much as the next leading spender and is the only advanced industrialized nation without publicly supported universal health care coverage.[52]

It may be that the improvements in health derived from commoditized health care products reaches a saturation point and that future gains will be much more dependent on improvements in prevention and health maintenance. Wilkinson (1996) also suggests that beyond a certain point economic growth as measured by Gross Domestic Product (GDP) and human health as measured by life expectancy are no longer related. What he finds when looking across countries that are beyond a growth threshold is that social factors such as equality and community cohesion are the primary correlates of health — even more so than individual factors such as smoking. Beyond the growth threshold commoditization and human health could be negatively correlated.[53]

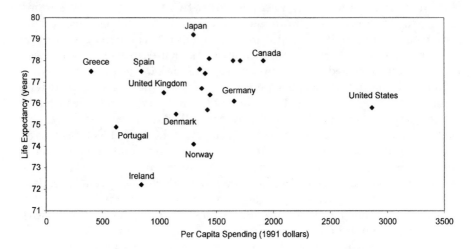

Figure 4.3   Health care spending and life expectancy for selected countries.

4.5   *Environmental pollution control and the 4Rs*

Like the health care industry, pollution control and environmental protection resist complete privatization because of their role as public goods, but they are still subject to the pressures of commoditization. The economics of environmental health and human health are driven by the same forces. As in health care, prevention may be preferable to treatment, but treatment involves products and services that can be readily marketed whereas prevention requires systems changes and the reform of production and consumption patterns that are not readily packaged solutions. As a result, waste reduction and pollution prevention remain underdeveloped and undercapitalized. Consider the following distinction between environmental products with high commodity potential and environmental processes and skills with lower commodity potential:

Environmental products (high commodity potential)
Remediation equipment
Energy-efficient appliances
Biological enzymes
Low-impact pesticides
Waste incinerators
Photovoltaic cells
Biomass fuels
Parks and zoos

Environmental system processes (low commodity potential)
Energy and materials conservation programs
Ecological design

> Watershed management
> Voluntary simplicity
> Community building and resource sharing
> Environmental education
> Waste-reduction programs
> Extended producer responsibility
> Habitat protection and conservation

Many analysts look to Nature for models of effective resource conservation. In natural systems there is no such thing as material waste; there are only by-products that become resources for another organism or process. By mimicking nature in the design of our production and waste management systems, energy and material efficiency can be enormously improved and wastes thereby dramatically reduced. Every step, from obtaining and processing raw materials to delivering final goods to repair and maintenance to final disposal, could be organized with the intention of minimizing the waste of energy and materials.[54]

There is a widely accepted formula for waste minimization, the 4Rs: Reduce, Reuse, Recycle, and Recover. These four approaches are typically depicted in order of effectiveness and priority (Figure 4.4).

- REDUCE the amount of waste produced: Reduce unnecessary packaging, improve energy and materials efficiency in both production and consumption, reduce the amount of material and energy required to provide any given service.
- REUSE materials: The waste material and energy that cannot be eliminated should be captured and reused in another part of the production

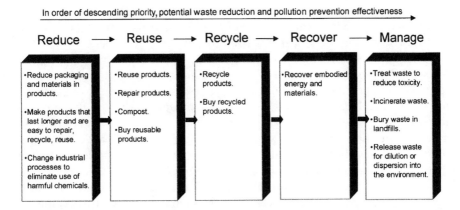

Figure 4.4   The highest priority and greatest potential effectiveness also has the least commodity potential. The greatest share of waste-related dollars goes to the lowest priority, i.e., waste management.

process. Waste can also be greatly reduced by repair and remanufacture of existing consumer products rather the manufacture and sale of new ones. For example, reconditioning a car to make it last for another 10 years requires 42% less energy and significantly less material than manufacturing a new car.[55]

- RECYCLE used materials: Materials that cannot be recovered for reuse, repair, or remanufacture should be recycled for use as raw materials somewhere else in the economy.
- RECOVER materials: Lastly, some component of the waste stream that is not suited for reuse as raw materials in other processes can sometimes be recovered as fuel for energy production or steam generation.

A formal policy on waste reduction was established in the U.S. by the 1976 Resource Conservation and Recovery Act, which focused originally on requiring industry to modify its production processes to encourage and facilitate recycling of raw materials. This approach, however, was considered too expensive by manufacturers, who successfully convinced the administration and Congress that RCRA should instead support landfill improvements and the construction of a new generation of solid waste incinerators.[56] Waste handling services and incineration have far greater commodity potential than does waste reduction, which in effect has negative commodity potential.

The priority ranking for the 4Rs is the exact opposite of the order of commodity potential. Although the 4Rs approach and prioritization are regularly advocated by environmental agencies in the U.S. and elsewhere, the actual practice reflects another set of priorities. Figure 4.5 shows the trends in expenditures for environmental regulation and monitoring, environmental research, and private pollution abatement services. The regulation, monitoring, and research categories, which include primarily services with mid-range or low commodity potential, have been declining, while the most highly commoditized sector of private pollution abatement services has been increasing.

A closer look at recycling as it is presently organized in the U.S. provides a good example of how commoditization distorts what is ostensibly an environmentally beneficial activity. The biggest problem with recycling is that it has never been fully integrated into a system of waste reduction that includes the entire life cycle of a product from raw materials extraction to production and consumption. In such a system recyclability would be designed and manufactured into products up front. Consumer products would contain materials that are meant to last beyond their first use, materials that are designed to be recycled. Repairability would also be designed and manufactured into the consumer goods we purchase so that repair would become less expensive than replacement.

Recycling as it is presently practiced has been organized to produce raw materials. This is consistent with the pattern we have been describing of privileging products over processes. In today's U.S. economy, it is

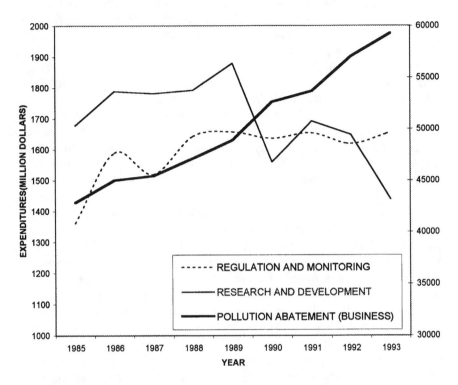

Figure 4.5 Pollution abatement and control expenditures in constant (1987) dollars: 1985–1993. (*Source: United States Statistical Abstracts, 1990, 1997.*)

considerably less expensive to throw away a damaged radio, for example, than it is to pay to have it repaired. As commoditizaton drives innovation, goods that were once repairable no longer are. For example, in recent years the sports-shoe industry has replaced rubber and leather with synthetic materials. As a result, the athletic shoes are now unrepairable.

Products are not designed for reuse, and therefore recycling programs must first transform the material they collect into usable materials again, a process that uses additional raw materials and energy and produces considerable pollution.[57] Even though science can produce materials that are easily recyclable, investments in this research and development are stymied because of the ambiguity of property rights over recycled materials in subsequent uses and liability for the environmental effects of post-consumer waste. As I have argued, the ease with which property rights can be established is the most significant characteristic of goods and services with high commodity potential.

One proposal for improving recycling and materials recovery is to require manufacturers to take back products, especially large consumer goods and appliances, once they have completed their service life. This would create significant incentives for manufacturers to design longevity and recyclability into their products. Companies facing such requirements or liability for the environmental effects of their waste may shift to leases rather than sales. This is already happening in some instances; for example, with some large carpet manufacturers who lease carpets to major institutional customers with the intention of recovering the worn carpets and using the materials in the production of new carpeting. Leasing is possible for a wide range of consumer goods (more on this in Chapter 8).

Since the 1970s the generation of waste products by industry in the U.S. has declined significantly. This is often pointed to as evidence that environmental problems can be solved by continued economic growth and that any efforts to slow growth to achieve environmental ends are counterproductive. However, only industrial waste has declined; consumer waste has risen precipitously. This is possible because a large percentage of the material thrown away by Americans was produced in other lands. For example, annual production of metals in the U.S. is more than 1.5 tons per capita (down from a maximum of close to 2 tonnes in the early 1970s). However, this decline merely reflects the fact that the U.S. is increasingly dependent on imported ores and metals. U.S. metal *consumption* is now more than 2.5 tons per capita.[58] In effect, U.S. environmental gains have been purchased by exporting industrial production and associated pollution elsewhere. Declines in industrial waste have not been matched by a decline in domestic refuse. *Per capita* waste has increased 58% between 1960 and 1995, meaning that not only are we producing more garbage because our population has grown, but also because each of us actually consumes more and discards more waste (Table 4.2).

Table 4.2 U.S. per Capita Refuse Output (pounds per day) in 1960, 1970, 1980, and 1995

WASTE MATERIAL	1960	1970	1980	1995
Total nonfood and nonyard wastes	1.65	2.25	2.58	3.26
Paper and paperboard	0.91	1.20	1.34	1.69
Glass	0.2	0.34	0.37	0.27
Metals	0.32	0.38	0.36	0.33
Plastics	0.01	0.08	0.19	0.39
Rubber and leather	0.06	0.09	0.10	0.12
Textiles	0.05	0.05	0.06	0.15
Wood	0.09	0.11	0.16	0.31
Other wastes	1.01	1.05	1.13	1.04
Food wastes	0.37	0.35	0.32	0.29
Yard wastes	0.61	0.63	0.67	0.61
Other	0.03	0.07	0.13	0.14
Total waste	2.66	3.30	3.71	4.30

Source: Franklin Associates, Ltd., Prairie Village, KS, Characterization of Municipal Solid Waste in the United States: 1995. Prepared for U.S. Environmental Protection Agency.

It is interesting to note that the presence of plastics, the most highly commoditizable material in the waste stream, has grown by a factor of 40 while metal and glass, materials with lower commodity potential, have increased only slightly. As a percentage of total waste, plastics have increased from less than 0.5% to nearly 10%, a 20-fold increase.

In many ways, plastics are the ultimate commodity materials. Almost infinitely malleable, they have greatly increased our capacity for standardization and packaging. Mostly synthesized from petroleum, plastics greatly expand the amount of fuel energy embodied in the products of everyday life. Their malleability makes them particularly well-suited for molding operations, which greatly increase the labor savings. Plastics are difficult to recycle and plastic products are almost impossible to repair. They have become a huge portion of the waste stream and degrade extremely slowly in the environment, if at all. The by-products of their oxidation, particularly when burnt, produce a considerable portion of the dioxins and other toxic pollutants in the air, water, and soil.[59]

4.6 Transportation

It makes sense to move people and goods as efficiently as possible. The efficiency of a transportation system can be measured in terms of the time required to travel between two locations or the amount of fuel or money expended per unit of distance traveled or the percentage of GDP used to support a nation's transportation system. Improving transportation efficiency in any of these ways requires organizing transportation services as a system whose purpose is the efficient movement of people and goods.

Commoditization distorts the transportation sector, as it does all other sectors, by influencing the direction of research and development and innovation and investment toward transportation-related products with the greatest commodity potential and away from the transportation system as a whole. In the case of the transportation system, automobiles and the physical infrastructure of roads and highways are the most commoditizable elements of the system. Previously (in Chapter 2) we have seen how the gas-powered internal combustion engine outcompeted both steam and electricity as the preferred method of powering a personal vehicle because of the greater commodity potential of fossil fuels. Now we will consider how the transportation systems built around the automobile perform in measures of system efficiency.

In terms of time required to travel between two locations in the center of major cities, street traffic today moves as or more slowly than horse-drawn transportation.[60] There has been little or no improvement in intercity travel speed since the early 1900s when urban trolleys were frequently linked with each other via intercity passenger rail. To drive from downtown in my home city of Syracuse, New York to nearby downtown Rochester takes from 90 minutes to 2 hours, depending on traffic and weather. In 1911 the scheduled travel time via intercity rail from Syracuse to Rochester was 70 minutes.

In Syracuse and Rochester and elsewhere throughout the U.S. a diverse mix of streetcars, rails, carriages and buses, some privately and some publicly owned, provided urban mass transportation in the early years of the twentieth century. In *The Geography of Nowhere: The Rise and Decline of America's Man-Made Landscape*, author James Howard Kunstler wrote:

> A civilization completely dependent on cars, as ours is now, was not inevitable. The automobile and the electric streetcar were invented and made commercially viable at roughly the same time: the period from 1890 to 1915. However, the automobile, a private mode of transport, was heavily subsidized with tax dollars early on, while the nation's streetcar system, a public mode of transport, had to operate as private companies, received no public funds, and were saddled with onerous regulations that made their survival economically implausible.[61]

Where government policy did not completely destroy streetcar systems, deliberate efforts by the automobile industry to put them out of business finished the job.[62]

Estimates of the cost of the U.S. transportation system vary depending on whether such externalities as the costs associated with air pollution and accidents are considered in the cost estimates or just the direct costs of vehicles, fuels, maintenance and infrastructure. Counting only direct costs, the U.S. Department of Transportation estimates the system costs around $1 trillion dollars or about 17% of U.S. GDP.[63] As a percentage of GDP this is nearly double what Japan spends on transportation and considerably more than most European countries spend. The more highly developed the passenger and urban mass transit systems are, the lower the overall costs of transportation, which is another way of saying that the more commoditization has distorted a nation's transportation system, the more transportation costs the nation's people and the less efficient it is.[64]

The costs of automobile accidents alone equal 8% of the U.S. Gross Domestic Product. In 1988, a study by the Urban Institute calculated that $71 billion were borne by citizens in out-of-pocket costs for accidents, another $46 billion in lost wages and household production, and $217 billion in pain, suffering, and diminished quality of life. This totals $334 billion in lost property, worktime, injuries, and deaths.[65] About 43,000 people are killed in cars each year and about 2 million sustain injuries serious enough to require hospitalization.

The negative impacts of over-reliance on personal automobiles and the underdevelopment of mass transit have been documented sufficiently elsewhere.[66] Safe, clean, and energy-efficient alternatives are possible and have been promoted, with some success, for decades: diversity of transportation options, safe bicycle lanes, car pooling, urban rail, intercity passenger rail, pedestrian malls and car-free zones, etc. The potential for advanced technology to revolutionize the efficiency of both public and private transportation is enormous.[67] Think of an electric dial-a-minibus that could respond to electronic requests to pick up passengers at nearby stops while computers continuously optimize the route of the buses in transit. Other options include electric cars, trolley cars, buses with bike racks, and on and on. No list can be complete. All of these innovations require thinking about transportation as a whole system and allocating resources and assessing costs in ways that improve overall system efficiencies. As we have seen, such systems approaches are consistently underdeveloped as a result of the economic distortions of commoditization.

4.7 Science and academia

Scientific and academic research can be categorized in ways that are roughly equivalent to the distinction we have made between product-oriented approaches with high commodity potential and process-oriented approaches with low commodity potential. In general, research with a focus at the level of the product (or system component) is favored over research into systems and their processes.

Advances are most often made in science through experiments that isolate a single variable or, at most, a small number of variables. Once experimental conditions are established, a variable of interest is manipulated and results are observed. In this way, knowledge is gained about component parts of reality. The more completely a science is capable of isolating its variables, the more is learned about them, the more precise one can be in speaking about them, the more predictable the behavior, and the more useful the knowledge becomes for engineering purposes. This is especially true for the hard sciences (as well as the hard sectors of the soft sciences).

Ecology studies the relationships and systems dynamics of Nature. Some ecologists study energy and nutrient flows, while others search for generalizable principles that help explain ecological patterns. Ecology rarely produces information that is readily transformable into products. Genetics, by comparison, studies the component building blocks of the living world. The results of genetic research are awaited by a vast and growing population of genetic engineers whose tinkering and product building is supported by industry's R&D infrastructure.

Before genetics became the most commoditizable of the sciences of Nature and thus captured the bulk of available R&D, there were other disciplines — silviculture and forest engineering, wildlife and fisheries biology and management, hydrology and hydrological engineering — that captured much of the investment in the sciences of Nature. At that time, these disciplines delivered the most commodity potential. But today the commodity potential of the traditional natural resources disciplines has been surpassed by biotechnology and genetic engineering, which promise to construct biological factories for fish, food, and fiber. Thus, in the hierarchy of commodity potential of the environmental sciences, genetics now prevails where natural resource management once did. Both are privileged over ecology.

Ecology, by studying natural systems as *systems,* fails to spin off a type of engineering that can maximize a single variable or optimize a single resource. Ecology uncovers unpleasant truths about commodity and profit-maximizing behavior. Whenever an ecosystem is managed to maximize the production of a single or a few economically valuable products, it invariably simplifies the system and the complex dynamics that produce the coveted resource to begin with. Ecology, and all systems sciences, stand in judgment against the very simplification and maximization of production efficiency that defines commoditization. It is therefore unsurprising to discover that ecology receives far less R&D support than the product-oriented disciplines, despite the widespread agreement that improved ecological understanding is critical for our future. With improved understanding of ecological processes and systems comes the potential to invent new technologies and ways of living in society that do not have the same simplifying effects as do technologies that maximize a single variable. We not only require a better understanding of ecological relationships and ecological systems, we also need policies that make us take account of these relationships (see Chapter 8).

Even within ecology, however, it is possible to place parts of the discipline on a scale of commodity potential. At the low end of commodity potential, ecology seeks improved understanding of ecological integrity, ecological dynamics and ecosystem health. At the high end, ecology seeks to elucidate broad ecological relationships that impact on the productivity and availability of commercially important products such as sport fish, game, forest products, etc. Often this ecology focuses on the development of simplified abstract ecological models designed to improve the capacity for managing ecosystems for increased productivity and sustainable yield. The distinction can also be understood by comparing alternative meanings of the terms ecosystem management.

4.7.1   An attempt at high commodity potential ecology

In October of 1970 President Richard Nixon signed a Congressional resolution authorizing $40 million for the National Science Foundation's participation in the International Biological Program, a grand enterprise that meant to model the world's major ecosystems. Planning for the IBP had been ongoing throughout the 1960s, and U.S. participation had begun in 1968 with minimal funding from the NSF. The new funding secured the future of the IBP in the U.S., significantly raised the status and funding structure of ecological research, and institutionalized the ecosystem concept in the mainstream of science policy.[68]

The impetus for the IBP was a recognition among ecologists as early as the 1950s that humanity was placing an increasingly heavy burden on the ecological systems of the Earth, and that resource and environmental problems were going to become significant problems in the near future.[69] In 1961, an organizational meeting in Amsterdam established three core areas that the IBP would investigate: human heredity, plant genetics and breeding, and the study of ecosystems that were likely to suffer negative effects of human activity.[70] In 1965 the National Academy of Sciences created a United States National Committee for the IBP. A meeting in Williamstown, Massachusetts in 1966 established a particularly American focus for the US/IBP: an ecosystems emphasis that involved "relatively well-funded, closely-linked, cooperative studies of whole ecosystem," what would become the "biome program."[71]

The IBP thus began with broad concerns that reflected a growing societal awareness of human impacts on nature and worries about the future capacity of humanity to utilize natural resources. Much of this concern was based on the fear of resource scarcities and food shortages resulting from an exponential growth in the human population.[72] The logic behind the three areas of emphasis established in Amsterdam was that this knowledge would be needed to manage and manipulate natural systems; that knowledge of human genetic diversity could help us understand how genetic adaptations to environmental conditions had helped human populations survive in the past; and that knowledge of plant genetics and breeding would contribute to higher sustained crop harvests. The idea behind IBP was clearly to put science to use in a big and coordinated way to solve emerging environmental problems that might one day threaten human well-being or survival, and to do so by allowing humans to have more direct control over the environment.

The US/IBP put a particularly American slant on the program by working off the development of ecosystem ecology to give it an ecosystems focus. Thus, the overall objective of the US/IBP was framed as an effort "to improve understanding of the interrelationships within and among ecosystems." This would be accomplished by:

- Formulating a basis for understanding the interactions of components of representative biological systems;
- Exploiting the understanding of biological systems to increase biological productivity;
- Providing bases for predicting the consequences of environmental stresses, both natural and man-made;
- Enhancing man's ability to manage natural resources;
- Advancing knowledge of man's genetic, physiological, and behavioral adaptations.[73]

Hearings of the Subcommittee on Science, Research and Development of the House of Representatives were held in 1967 to gain Congressional support for the IBP. Subsequent to the hearing the subcommittee issued a report urging funding for the US/IBP.[74] The report relied heavily on the testimony it had heard concerning the grave environmental dangers facing humanity and the necessity of creating a science-policy infrastructure to begin to address them. As the subcommittee concluded, the justification for the IBP was that it would bring science to bear on "one of the most crucial situations to face this or any civilization — the immediate or near potential of man to damage, perhaps beyond repair, the ecological system of the planet on which all life depends."[75]

The testimony before the subcommittee indicated that the US/IBP would make significant contributions in developing a theoretical ecology that would have predictive capacities then lacking, and advance knowledge of ecosystems in order to more directly control and effectively manage natural resources.[76] Some witnesses expressed a larger vision of what the ultimate ends of the IBP might be. Stanley Cain, the Assistant Secretary of the Department of the Interior for Fish, Wildlife and Parks, argued that the:

> IBP is really the world's first organized effort to face up to this class of vital problems that deal with the limits of natural productivity in various ecological systems. The possibilities of human management of such systems extend before us new frontiers that can be reached if we develop and apply ecological knowledge.[77]

The ambition to manage ecosystems as systems was an extension of the then-popular cybernetic systems theory in the sciences. The notion of cybernetic control of society, in which society is conceived of as a closed system that the controller can stand outside of and manipulate, developed a following among some policy-makers and bureaucracies in the 1960s. The utilization of the concept in the context of the environment would have resonated with the "more general cultural repertoire" than Congresspersons were acquainted with at the time.[78]

The US/IBP was distinguished by the biome studies, which divided the country into five large ecological regions that were to be studied as whole ecosystems. Seventy percent of the funding allocated to the US/IBP went to the biome studies. The central goal of the biome program was to model ecosystems as comprehensively as possible. The emphasis on modeling points up the quantitative, abstract, and ultimately commoditizable posture taken by the IBP effort. Though a basic science program, the IBP was expected to pay off in terms of providing resource managers with tools to help them maintain ecological functions in order to maintain the steady flow of products from nature.

Though the sciences involved in the IBP — ecology, mathematics, computer science — differed from the traditional resource management sciences such as silviculture, tree physiology, and so forth, the cybernetic ideal and the traditional managed landscape ideal were strikingly compatible on their face: both sought to impose rational, steering control over the natural world. Instead, ecosystem ecology revealed complexity, relationship, and interdependence, the very qualities that traditional resource science sought to deny or abolish on the landscape. Traditional resource sciences assumed that they could disaggregate those parts of the natural world that yielded the greatest profits from those that did not, and that were considered irrelevant and not interesting subjects for research. Resource scientists assumed that nature could be made to resemble a machine, and that with sufficient manipulations and inputs the natural landscape could be reduced to a resource farm.

Ecosystem ecology, on the other hand, begins with the assumption that ecosystems *are* webs of complex relationships. Though they had started out with other goals in mind, IBP researchers discovered that ecosystems were even more complex than they had thought and that simplified models of ecosystems only began to capture their structural and functional relationships. Consequently, and without the approval of funding institutions, they abandoned the cybernetic ideal.

There are two key lessons from the IBP experience. The first is that the ecological complexities of nature defy attempts to simplify them. Even a "big science" project like the IBP, with many millions of dollars at stake, could not in the end bend the science of ecology to serve the forces of commoditization. But the second lesson is that those forces should be expected to at least *try* to use virtually all of the sciences and all knowledge to support their basic goal, the maintenance of the growth economy. To the extent that ecology can show that ecosystems will not play by our economic rules, it stands as perhaps the most promising of all the sciences. Understood and used correctly, ecology will help us reform our own practices and institutions to be more adaptive and consistent with the fundamental complexities of ecological systems, to live more lightly on the Earth rather than to put all our resources and attention on changing the basic structures of nature to produce commodities.

4.8   Electricity sector

The previous examples should provide a clear understanding of the mechanism behind commoditization, along with its environmental consequences. In the electricity-generation sector the story is the same. We have developed a large appetite for electricity to fuel our growing economy, and we have several options capable of providing this service. As with each sector discussed so far, research and development investments are allocated to the many competing commodities based on their present and historical abilities to prosper in a highly commoditized economy. The commodities that attract the greatest share of research and development investment naturally become the overwhelming sources of electricity. Unfortunately, as is typical in so many sectors of the economy, these high commodity potential sources also produce the greatest environmental impact.

My colleagues and I attempted to quantify the mechanism and environmental consequences of commoditization using six commodities from the electricity-generation sector. The analysis is based on three relationships. First, we examined the relationship between commodity potential and research and development investment to demonstrate the fact that goods and services with high commodity potential tend to attract the lion's share of R&D investments (Figure 4.6). Second, we examined the relationship between commodity potential and percent use to demonstrate the fact that these high commodity potential forms of electricity supported by the lions share of R&D funding tend to become the dominant sources of electricity (Figure 4.7). Finally, we examined the relationship between commodity potential and environmental impact to demonstrate that the commodities which receive the most R&D and become the dominant sources of electricity also cause the most environmental damage (Figure 4.8). The data necessary for such an analysis come from a variety of sources (Table 4.3). Percentage use (% Use) simply represents the consumption of the particular commodity divided by the sector total. Environmental Impact

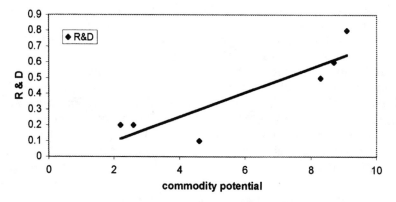

Figure 4.6   The relationship between commodity potential and electricity R&D for each of the six commodities for the electricity-generation sector ($p = 0.019$, $r^2 = 0.79$).

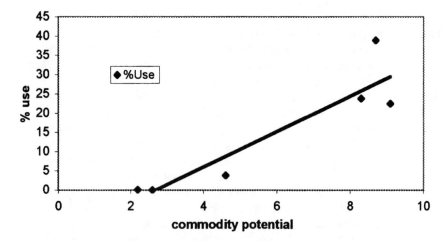

Figure 4.7   The relationship between percent use and commodity potential for each of the six commodities from the electricity-generation sector ($p = 0.11$, $r^2 = 0.84$).

Table 4.3 The Data Represent Values Averaged Over the Time Period from 1990 to 1996.

Commodity	% Use	Research and Development	Environmental Impact	Commodity Potential
coal	22.5	0.8	5.8	9.1
oil	38.9	0.6	2.3	8.7
natural gas	23.8	0.5	2.0	8.3
hydro	3.8	0.1	0.8	4.6
wind	0.04	0.2	0.1	2.6
photovoltaics	0.1	0.2	0.2	2.2

% Use data comes from the DOE's web site: http://www.eia.doe.gov. R&D investment data comes from McKinney and Schoch, 1998, "Environmental Science: Systems and Solutions," p. 238. Environmental impact has two components: Land Use Data are from Brown et al., 1995, "State of the World." Emissions data are from a special feature article on the DOE's web site: http://www.eia.doe.gov/environment/ by John Carlin entitled "Environmental externalities in electric power markets: Acid rain, urban ozone and climate change." The data representing Commodity Potential are explained in the text.

(EI) represents a composite index including land use and emissions. The overall environmental impact of each of the commodities discussed was calculated using the following equation:

$$EI = (\text{Land Use} \times 0.0001) + [SO_2] + [NOx] + ([PM10] \times 10)$$
$$+ ([VOCs] \times 10) + ([CO_2] \times 0.001)$$

where land use is the amount of land necessary to produce a quadrillion BTUs of electricity, $[SO_2]$ is the concentration of sulfur dioxide, $[NOx]$ is the concentration of nitrogen oxides, $[PM10]$ is the concentration of particulate matter with diameter less than ten microns, $[VOCs]$ is the concentration of volatile organic compounds, and $[CO_2]$ is the concentration of carbon dioxide. Research and development expenditures represent the global dollars invested in each of the commodities discussed. Finally, commodity potential is a qualitative estimate. Each commodity is assigned a relative score of 1 to 10 for each of the criteria referred to in Chapter 2 (Table 2.2). The scores are then averaged to represent a single value for each commodity. Although the choice of scores for each criteria may be arbitrary, the relative scores are quite consistent. Therefore, this seems to be a reasonable first approximation of the commodity potential of several competing goods within the electricity-generation sector.

These data and the analyses based upon them are preliminary. We are currently working to improve the quality of our data for the electricity-generation sector, along with many of the other economic sectors referred to throughout this book. The percent use data for the electricity-generation sector are relatively easy to obtain and therefore require no additional comment. Environmental impact data are also widely available, but it is difficult

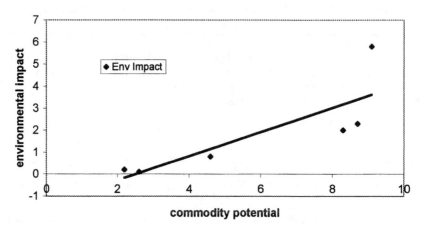

Figure 4.8   The relationship between environmental impact and commodity potential for each of the six commodities from the electricity-generation sector ($p = 0.048$, $r^2 = 0.66$).

to incorporate the data into a single integrated measure of environmental impact. Our measure, including emissions and land use, is a good start, but other factors such as energy return on investment should be included in future measures. The data to be integrated into a single measure of environmental impact will likely vary across economic sectors, suggesting the need for an overall, comprehensive measure of environmental impact appropriate for all sectors.

Research and development expenditures are obviously a key component to analyses of this type. Surprisingly, accurate estimates of these expenditures can be difficult to obtain depending on the economic sector under analysis. For the electricity-generation sector we have settled for global R&D investment data, and are currently searching for more appropriate and reliable national numbers. Finally, commodity potential, which is arguably the most important quantitative measure for analyses of this type, is also the most difficult to obtain. In future analyses across economic sectors the qualitative estimate we describe for the electricity generation sector must be made quantitative. We feel the best approach to this difficult task is to develop quantitative indicators for each of the categories referred to in Chapter 2 (Table 2.2), such as the number of patents issued for the "ability to assign and protect property rights" category, or total package delivery miles for the "degree of mobility" category.

Our preliminary analysis of the electricity-generation sector should be taken more as an example of how such analyses may be conducted rather than hard evidence supporting the theory of commoditization. However, despite (or perhaps because of) the crudeness of the initial data sets (Table 4.3) the results (Figures 4.6, 4.7, 4.8) support the mechanism and environmental effects of the notion of commoditization described in this book. Figure 4.6 demonstrates that R&D investments can explain 79% of the variance in commodity potential for the goods in the electricity-generation sector. Figure 4.7 demonstrates that this commodity potential can explain 84% of the variability in percentage use of the six electricity alternatives. Finally, Figure 4.8 demonstrates that commodity potential can also explain 66% of the variance in environmental impact attributable to the six electricity alternatives. The statistically significant and positive correlations for each of these relationships suggest that, consistent with the theory of commoditization, those goods in the electricity generation sector with the highest commodity potential receive the most R&D investments, tend to be used most often, and cause the greatest environmental impact.

4.9   Conclusion

Commoditization is the most ubiquitous economic phenomenon in the modern economy, yet, because of its ubiquity, it remains largely unidentified and transparent to policy-makers and the average person alike. The process of commoditization, and the policies that maintain it, are built into our assumptions about the way the world works. In other words, they are naturalized and seem inevitable rather than socially constructed. When one begins to examine discrete economic sectors and policies, however, commoditization can be shown in greater relief, and its pervasive spread appears to be more a result of the structural conditions of our particular economy than economy in general. The purpose of this chapter has been to provide examples of commoditization in familiar economic activities in order to lay the foundations of future research and for developing specific policy alternatives to redress the economic imbalances that commoditization has produced.

The examples in this chapter demonstrate that commoditization is no mere occasional or even strictly commercial phenomenon in the sense of being limited to private industry. Rather, it is a fundamental development process in a society that has become dependent on rapid and unending economic growth. We no longer ask whether a product or a service is good for society as such, but whether it is profitable and can be used to produce more products or services in the future. In our system, commoditization and the economic growth that results from it have become self-justifying. The social good has been replaced by the economic good, and rather than having the economy be at the service of society, society is now at the service of the economy. The assumption of liberal theorists has been that by some sort of invisible magic this arrangement actually results in the greatest social good that is possible. This assumption is no longer tenable in a world of diminishing resources, growing pollution, and growing populations. It is time we seriously reconsider our assumptions.

Changes in the economic system need not be made wholesale or suddenly. Fortunately, immediate alternatives are often available to the most highly commoditized options for many areas of need satisfaction. Alternative, low-input agriculture has managed to survive despite the massive investments that have been made in input-intensive industrial agriculture. To the extent that organic farming has made compromises with the modern economy by developing commoditized approaches to producing organic foods, it has also maintained and even expanded knowledge and skills that can be used to more fully develop its potential. Alternatives to commodity-intensive health care, transportation, and knowledge production have also survived, as has the commitment on the part of a small number of people to keep those options viable. In order for them to spread, however, a more general consensus will have to emerge among the public. Ultimately, it will be the public that has to demand changes in the economic structure that make LCP options competitive with HCP options across the spectrum of economic activity. Exactly how this could come about will be discussed in other chapters of this book.

Notes

1. Landsberg, H.H., Materials: some recent trends and issues, *Science,* February 20, 1976, 637–641; See also Max-Neef, M., Economic growth and quality of life — a threshhold hypothesis, *Ecological Economics,* 15(2), 115, 1995.
2. Organic Farming Research Foundation, Frequently Asked Questions About Organic Farming, at http://www.ofrf.org. July 17, 1998.
3. U.S. Department of Agriculture, Agriculture FACT BOOK 98, http://www.usda.gov/news/pubs/fbook98/chart1.htm, p. 10.
4. U.S. Department of Agriculture FACT BOOK 98.
5. Fuglie, K., Ballenger, N., Day, K., Klotz, C., Ollinger, M., Reilly, J., Vasavada, U., Yee, J., *Agricultural Research and Development: Public and Private Investments under Alternative Markets and Institutions,* U.S. Department of Agriculture, Agricultural Economic Report Number 735, Washington, D.C., May 1996.
6. Edwards, R., Tomorrow's bitter harvest, *New Scientist,* 151(2043), August 17, 1996, 14.
7. Diamond, J., *Guns, Germs and Steel,* W. W. Norton, New York, 1997.
8. For a good summary see "The Importance of Indigenous Knowledge," at the the Smithsonian Institute Website http://www.si.edu/biodiversity/aislynn1.htm, 5/7/99.
9. Moser, A., Ecological engineering and bioprocessing, in: *Eco-restructuring: Implications for Sustainable Development,* Ayres, R.U. (Ed.), United Nations University Press, Tokyo, 1998, pp. 77–108.
10. Hecht, S., Indigenous soil management, in: *Let Farmers Judge,* Hiemstra, W., Reijntjes, C., and van der Werf, E. (Eds.), Intermediate Technology Publications, London, 1992, p. 137.
11. Hecht. S. and Cockburn. A., *The Fate of the Forest,* Harper Perennial, New York, 1990, p. 48. See also Norgaard, R., *Development Betrayed,* Routledge, London, 1994, pp. 104–120.
12. National Research Council, *Sustainable Agriculture and the Environment in the Humid Tropics,* National Academy Press, Washington, D.C., 1993.
13. Wohlmeyer, H., Agro-eco-restructuring: potential for sustainability, in: *Eco-Restructuring: Implications for Sustainable Development,* Ayres, R.U. (Ed.), United Nations University Press, Tokyo, 1998, p. 291.

14. Murdoch, W., World hunger and population, in *Agroecology,* Carroll, C.R., Vandermeer, J.H., and Rossett, P.M. (Eds.), McGraw-Hill, New York, 1990, pp. 3–20.

15. Weatherford, J.M., *Indian Givers: How the Indians of the Americas Transformed the World*, 1st ed., Crown Publishers, New York, 1988, p. 83.

16. Pimentel, D., Harvey, C., Resosudarmo, P., Sinclair, K., Kurtz, D., McNair, M., Crist, S., Spritz, L., Fitton, L., Saffouri, R., and Blair, R., Environmental and economic costs of soil erosion and conservation benefits, *Science 267*, 111, 1995.

17. Matson, P.A., Parton, W.J., Power, A.G., and Swift, M.J., Agricultural intensification and ecosystem properties, *Science*, 277, 1997, pp. 504–509.

18. Gallopin, G.C., Eco-restructuring tropical land use, in *Eco-Restructuring: Implications for Sustainable Development*, Ayres, R.U. (Ed.), United Nations University Press, Tokyo, 1998, p. 317.

19. U.S. Department of Agriculture Economic Research Service, Change in Livestock Production, 1969-1992, AER-754, August 1997 at http://www.econ.ag.gov/epubs/htmlsum/are754.htm, 5/11/99.

20. Pimentel, D. and Pimentel, M. (Eds.), *Food, Energy and Society*, University Press of Colorado, Niwot, CO, 1996, p. 168.

21. Pimentel and Pimentel (1996), p. 8.

22. Hall, C.A.S. and Hall, M.H.P., The efficiency of land and energy use in tropical economies and agriculture, *Agriculture, Ecosystems and Environment*, 46, 1993, pp. 1–30.

23. Ophuls, W., *Requiem for Modern Politics*, Westview, Boulder, CO, 1997, p. 100.

24. Wohlmeyer, H. (1998), p. 292.

25. Wohlmeyer, H. (1998). See also Anderson, M., The life cycle of alternative agricultural research, *American Journal of Alternative Agriculture*, 10(1), 1995.

26. Lipson, M., *Searching for the O-Word*, Organic Farming Research Foundation, Santa Cruz, CA, 1997.

27. Anderson (1995).

28. Wohlmeyer (1998).

29. Warick, J., Cultivating farms to soak up a greenhouse gas, *Washington Post*, 1998, p. A03.

30. Brown, L.R., Gardner, G., and Halwell, B., *Beyond Malthus: Sixteen Dimensions of the Population Problem, Worldwatch Paper 143,* Worldwatch Institute, Washington, D.C., 1998, p. 31.

31. Ehrlich, P.R., Ehrlich, A.H., and Daily, G.C., *The Stork and the Plow,* G.P. Putnam, New York, 1995, p. 171.

32. Lilliston, L. and Cummins, R., Organic versus "organic": the corruption of a label, *The Ecologist*, July/August 1998. See also Tripp, R. and van der Heide, W., The erosion of crop genetic diversity and uncertainties, *Natural Resource Perspectives*, March 7, 1996, Overseas Development Institute, London, 1996. See also the National Organic Program webpage, http://www.ams.usda.gov/nop/rule/summ1.htm, 2/17/99.

33. Lipson, *Searching for the O-Word.*

34. Lilliston and Cummins, "Organic versus 'organic.'" See also U.S. Department of Agriculture "Questions and Answers about the National Organic Program Proposed Rule" at http://www.ams.usda.gov/nop/rule/Resource%20Material%20Files/qs&as.htm. 5/18/99.

35. McKinney, M. and Schoch, R., *Environmental Science: Systems and Solutions,* Jones and Bartlett, Boston, 1998.

36. For a summary of the case against biotechnology see Lappé, M. and Bailey, B., *Against the Grain: Biotechnology and the Corporate Takeover of Your Food*, Common Courage Press, Monroe, ME, 1998.
37. Paoletti, M.G. and Pimentel, D., Genetic engineering in agriculture and the environment, *Bioscience*, 46(9), 1996, 665–673.
38. Zuger, A., Fever pitch: getting doctors to prescribe is big business, *New York Times*, January 11, 1999, pp. A1, A13.
39. LaRosa, J.H. and Kiefhaber, A., Cost analysis and workplace health promotion programs, *Occupational Health Nursing*, 33, 234, 1985. See also Pender, N., *Health Promotion in Nursing Practice*, Appleton & Lange, New York, 1996, p. 299.
40. Farquhar, J.W., Case for dissemination research in health promotion and disease prevention, *Canadian Journal of Public Health*, 87(2), S44, Nov-Dec, 1996.
41. PhRMA, 49 New Treatments Added to Nation's Medicine Chest; Companies to Invest $20.6 Billion in Research, Web publication http://www.phrma.org/charts/nda%5Fi97.html, 10/30/98.
42. Lamptey, P. and Cates, W., An ounce of prevention worth a million lives, *Network*, 17(2), Family Health International, Winter 1977.
43. U.S. Department of Health and Human Services, *Clinton Administration Record on HIV/AIDS, Fact Sheet, March 12, 1996*, http://www.hhs.gov/news/press/1998pres/980312b.html. 3/9/99.
44. CIS Statistical Universe, No. 154, National Health Expenditures, by object: 1980 to 1995, http://web.lexis-nexis.com/statuniv/attachment/attachment.gif?_butinfo=, 4/12/99.
45. CIS Statistical Universe, No. 173 Health Maintenance Organizations (HMOs): 1980 to 1995, http://web.lexis-nexis.com/statuniv/attachment/attachment.gif?_butinfo=, 4/12/99.
46. Gorz, A., *Ecology as Politics*, South End, Boston, 1980, p. 183.
47. Hall, C.A.S., Pontius, G., Ko, J.-Y., and Coleman, L., The environmental impact of having a baby in the United States, *Population and Environment*, 15, 1994, 505–523.
48. Hynes, P.H., *Taking Population out of the Equation: Reformulating I=PAT*, Institute on Women and Technology, North Amherst, MA, 1993. See also Jessica Matthews of the World Resources Institute quoted in Ayres, R. U., Eco-restructuring: the transition, in *Eco-Restructuring: Implications for Sustainable Development*, Ayres, R.U. (Ed.), United Nations University Press, Tokyo, 1998, p. 21.
49. Hynes (1993); Matthews, in Ayres (1998).
50. Tainter, J.A., *The Collapse of Complex Societies*, Cambridge University Press, Cambridge, 1988.
51. Stewart, C.T., *Healthy, Wealthy or Wise? Issues in American Health Care Policy*, M.E. Sharpe, Armonk, NY, 1995, p. 11.
52. Stewart (1995), p. 223.
53. Wilkinson, R.G., *Unhealthy Societies: The Afflictions of Inequality*, Routledge, New York, 1996.
54. U.S. Environmental Protection Agency, Extended Product Responsibility: A New Principle for Product-Oriented Pollution Prevention, June 1997, Website report at: http://www.epa.gov/epaoswer/non-hw/reduce/epr/epr.htm.

55. Stahel, W. and Jackson, T., Optimal utilization and durability, in *Clean Production Strategies*, Jackson, T. (Ed.), Lewis Publishers [for the Stockholm Environment Institute], London, 1993.

56. Gould, K.A., Schnaiberg, A., and Weinberg, A.S., *Local Environmental Struggles: Citizen Activism in the Treadmill of Production*, Cambridge University Press, New York, 1996, p. 146.

57. McDonough, W. and Braungart, M., The NEXT Industrial Revolution, *Atlantic Monthly*, October 1998.

58. Rohatgi, P., Rohatgi, K., and Ayres, R., Materials futures: pollution prevention, recycling and improved functionality, in *Eco-Restructuring: Implications for Sustainable Development*, Ayres, R.U. (Ed.), United Nations University Press, Tokyo, 1998, p. 111.

59. DePinto, J., Manno, J., Milligan, M., Guzman, R., Honan, J., and Lopez Lemus, L., *Proposed Chlorine Sunsetting in the Great Lakes Basin: Policy Implications for New York State*, Donald W. Rennie Memorial Monograph Series, No. 11, State University of New York at Buffalo, Buffalo, NY, 1998.

60. Roodman, D.M., *The Natural Wealth of Nations*, W.W. Norton, New York, 1998, p. 89.

61. Kunstler, J.H. *The Geography of Nowhere: The Rise and Decline of America's Man-Made Landscape*, Simon & Schuster, NY, 1993, p. 86.

62. Kunstler (1993), p. 91.

63. U.S. Department of Transportation, *Transportation Statistics 1994*, U.S. Department of Transportation, Bureau of Transportation Statistics, Washington, D.C., 1994, pp. 4–5.

64. Kay, J.H., *Asphalt Nation*, Crown Books, New York, 1997. See also Hook, W., Counting on Cars, Counting out People, Institute for Transportation Development Policy Paper, New York, 1994, p. 28; Replogle, M., Improving access for the poor in urban areas, *Appropriate Technology*, 20(1), 1993, 21–23; World Resources Institute, *The Going Rate: What It Really Costs to Drive*, World Resources Institute, Washington, D.C., 1992; Conservation Law Foundation, *Road Kill: How Solo Driving Runs Down the Economy*, The Conservation Law Foundation, Boston, May, 1994, p. 7.

65. Miller, T., *The Costs of Highway Crashes*, The Urban Institute, Washington, D.C., October 1991.

66. Kay (1997).

67. Lovins, A. and Lovins, H., Reinventing the wheels, *The Atlantic Monthly*, January 1995.

68. Kwa, C., Representations of Nature mediating between ecology and science policy: the case of the International Biological Programme, *Social Studies of Science*, 17, 1987.

69. McIntosh, R.P., *Ecology Since 1900: History of American Ecology*, Arno Press, New York, 1977, p. 363.

70. Golley, F., *A History of the Ecosystem Concept in Ecology*, Yale University Press, New Haven, CT, 1993, p. 111.

71. McIntosh (1977), p. 363.

72. McIntosh (1977).

73. National Research Council, *U.S. Participation in the International Biological Program*, National Academy of Sciences, Washington, D.C., 1974, p. 1–2.

74. Subcommittee on Science, Research, and Development, *The International Biological Program, Its Meaning and Needs,* U.S. House of Representatives, Washington, D.C., 1968.
75. Subcommittee on Science, Research, and Development (1968), p. 2.
76. National Research Council (1974), p. 12.
77. National Research Council (1974), p. 54.
78. Kwa (1987), pp. 425–426.

chapter five

Systematic oppression

Contents

5.1   Introduction

This chapter discusses some of the most emotionally charged concepts in the English language: racism, sexism, and oppression. You may feel criticized for your place in the hierarchies of systems of privilege and may want to argue with the text as you read along. I ask that you consider these ideas in the spirit with which they are proposed, as a way of understanding the effects of commoditization on society. Until now we have talked about the privileging of certain things over others in the allocation of resources. We now turn to how commoditization affects the social status of those who participate most fully in the commoditized economy compared to those who, for one reason or another, do not. These effects, as it turns out, neatly parallel the effects of various *"isms."* It is interesting and informative to understand how and why.

5.2   Linking oppression and commoditization

Commoditization creates a hierarchy of value where what is most readily marketable receives the largest share of society's resources. As a result, what gets pruned from our existence is very much of what matters in life. Lots of people who rely on traditional ways of life get pruned. People who have less of the qualities the system "wants" get pruned. There is a real matter of oppression resulting from the economic evolution that commoditization drives. Certain groups of people, because of historic circumstances, cultural resistance to the values of commoditization, and/or because of their close ties to sectors of the economy with low commodity potential, find themselves systematically undervalued in terms of the share of society's resources that flow to them and the things they value. This inequality in access to resources and power is the very definition of economic and, in part, political oppression. This chapter proposes a way of understanding the links between commoditization and oppression.

Those groups of people who have borne the greatest costs and gained the least from the benefits of commoditization — indigenous people and other people of color, women, people with disabilities, and everyone who has only their labor to sell (working-class people) — are the targets of racism, sexism, classism, and disability oppression. Although these "isms" are usually thought of in terms of personal prejudices, their roots and reinforcement lie in part in the economic biases of commoditization. Unless the economic forces of commoditization are addressed, the effects of this oppression will not be redressed.

There are two separate issues regarding the connection between economics and oppression, and we should be clear about the distinction. First there is the very real and very devastating inequality in access to resources. This inequality is growing rapidly more severe. At a recent conference at New York University on asset ownership, economist Edward Wolf points out that

- Adjusting for inflation, the net worth of the median American household fell 10% between 1989 and 1997, declining from $54,600 to $49,900. The net worth of the top 1% is now 2.4 times greater than the poorest 80%.
- The modest net worth of white families is 8 times that of African Americans and 12 times that of Hispanics. The median financial wealth of African Americans (net worth less home equity) is $200 (1% of the $18,000 for whites) while that of Hispanics is zero.
- Between 1983 and 1995, the bottom 40% of households lost 80% of their net worth. The middle fifth lost 11%. By 1995, 18.5% of households had zero or negative net worth (an average –$5,600, down from –$3,000 in 1983).[1]

This kind of inequality can and should be addressed by deliberate social intervention in the economy to achieve more equitable distribution of wealth. This is not, however, the main focus of this chapter. Commoditization creates in many ways a deeper and even more profound oppression than inequality. This assumption challenges most of the current efforts to eliminate oppression, which focus on inclusion without necessarily changing the economic patterns of biases associated with commoditization.

While the noncommodity ties of care and connection suffer from lack of resources, in every oppressed group there exists a small but growing middle class that is blessed with or had forebearers blessed with talents, ambitions, or economic luck that are "fit" for surviving or thriving in a commoditized economy. Because of limited advances made in decreasing overt race- and gender-based exclusion from skilled production jobs, markets, and capital in the past half-century (particularly in Western industrialized countries and in the areas of prosperity in the Third World), there are always some opportunities for success among individuals in oppressed groups. Many succeed, and their success is often presented as case arguments against the charge of racial or class bias. This is true both for oppressed groups within Western industrialized nations and for those in the former colonial territories. "Successful" individuals from oppressed groups experience at the same time the impoverishment of the common goods that are the lifeblood of their communities. Their privileged position and/or wealth allows them to purchase the ever-increasing number of commodity replacements for common goods (security devices instead of community spirit, professional services instead of extended families, etc.).

This imbalance, personally and painfully felt, simply mirrors the fundamental imbalance created by commoditization pressures throughout society. It is felt most acutely by members of oppressed groups exactly because the system that has been responsible for their oppression appears to be the only practical alternative for promoting individual and social welfare. Since the system appears as perfectly neutral and open for everyone, those who have been left out experience it as their own fault. And as more and more spheres of human needs and wants become dominated by commodity satisfactions, participation or membership in social life itself depends on ability to participate in the market. At the same time, certain aspects of oppressed people's culture is commoditized by the entertainment and sports industries so, for example, opportunities to experience the rhythms and power of African American culture can be purchased for the price of a CD or a ticket to a basketball game, while images of Native American culture are commoditized and marketed by Disney and Hollywood. This is an example of how commoditization pressures operate to simplify and stereotype by highlighting and marketing the most commoditizable aspects of any culture or group and ignoring or suppressing those aspects that resist commoditization.

Those whose energies focus on the particular, the local, the community are expending energies on nonrewarding services because by being local and particular they are less suitable as commodities. Not surprisingly the community-building tasks, which are inherently local and particular, are selected out as the people who devote themselves to those tasks find it increasingly difficult to sustain themselves. The geniuses of community organizing and caretaking easily fall victim to self-deprecation, as the primary vehicle for social valuation in a commoditized society, purchasing power, is systematically denied them.

Thus the oppression is internalized as belief in the lesser value of the oppressed group or individual, or it simmers as a deeply felt resentment against faceless and nameless oppressors. This difficulty with naming oppression is realistic: it results from the fact that the dynamics of oppression are simply the flip-side of the dynamics of commoditization, which are embedded in the system and are not necessarily the result of individual decisions to oppress other human beings. Although oppressive actions, attitudes, and behavior continue to be widely exhibited and experienced in people's daily lives, the brute force of such oppression has been diminished and isolated by changes in public policy and personal mores in many parts of the world. In one way, this easing of the personal relationships of oppressor and oppressed can be understood as an outcome of the success of commoditization. Economic forces have become so thoroughly enmeshed in the activities of daily living that their injustices, particularly the destruction of the substance of community life, become invisible as purposeful acts of oppression and instead are experienced as personal problems engendered by modern life itself.

In the present circumstances we almost all contribute to our own oppression and the impoverishment of our communities. We have little choice. A rational economic actor will put her time and attention toward activities and things likely to be most rewarding to her. When rewards are allocated according to commoditized criteria, people rationally make choices to give their time and attention to the most commoditized option before them. People who for whatever reason put their creative intelligence and time and attention toward the noncommoditizable activities of community building — or soil building, or developing a deep personal knowledge and understanding of local ecosystems, or working at the bedsides of people who are ill, or playing with children, or making things from local materials, or working with community-supported agriculture and countless other things that people do to make lives better — will suffer both financially and in terms of personal self-esteem when their good work is systematically denied the resources of society.

5.3   How to build community

A neighborhood becomes a community over time. It is built and maintained with products and services like bricks and mortar for the homes, schools, and businesses; the goods that fill the shops; and the tools, equipment, and skills that keep the neighborhood clean, repaired, and maintained. All these things matter a great deal to the quality of life in a neighborhood. But other things that cannot be easily bought and sold matter as well, perhaps even more. The time spent with neighbors, the sharing of childcare, the watching out for each other's homes and children, the communal enforcement of neighborhood standards, the walks, the parties, the births, weddings, and funerals. The Syracuse Cultural Workers, a nonprofit progressive art publisher and distributor located in my hometown, produced a poster on "How to Build a Community" (Figure 5.1). There is very little on the list that one can buy; it is full of things to *do*.[2]

Figure 5.1   How to Build Community. © 1998. Syracuse Cultural Workers. With permission.

In the 1950s and 1960s, the U.S. undertook a massive program of "urban renewal." In effect it was a prototypical process of commoditization imposed on entire communities. The downtown areas of many cities were aging; their infrastructure was worn, their streets too narrow for the expected increases in automobile traffic. The need for major reinvestment in cities was clear. What form urban renewal would take was determined largely by the politics and economics of the construction industry and the calculus of "efficiency." It was more "rational," according to these calculations, to level large sections of inner cities rather than to work with the people of inner city communities and invest in improved mass transportation and refurbishing and revitalizing the old neighborhoods. Skyscrapers and highways went up where neighborhoods and communities had been. Many people lost their homes; some were relocated into new public housing units designed with the same distorted notions of efficiency that had motivated the urban renewal projects. The middle class segments of the former community left, draining inner cities of business and capital. By the logic of economic growth, urban renewal was a success. Replacing inner city communities with highways into and out of downtown made the massive growth in suburbs possible, which in turn became a prime mover of U.S. economic growth. Former "slums" and "ghettoes" were replaced with attractive housing for young professionals. With businesses filling the new office towers and well-to-do tenants moving into the condominiums, property tax revenues rose.[3]

Next to these gains, the losses remained mostly hidden and untallied. The costs were borne by the people who lost contact with their neighbors and friends, lost their informal networks of support, lost the time they had invested in building community ties, which now had to be rebuilt from scratch.[4] Nobody had figured these losses into the calculations that determined the economic costs and benefits of urban renewal. Unlike bricks and mortar, these values cannot be bought and sold, or easily recreated. They are not, in the language of our analysis, readily portable; property rights can't be assigned to them, they are not centralizable, they are not commoditizable, and they can be readily forsaken in a system of social decision-making dominated by the rationale of efficiency maximization. There have been countless post-mortems of the devastating impact of urban renewal on U.S. cities, but little note has been taken of the underlying dynamics in which the stuff of highways and towers are the commodities of economic growth where much of what constitutes the life in city neighborhoods is not.

It is interesting to look at how even the ways cities are studied suffers the same kind of commoditization effect. In advance of urban renewal, large amounts of money were invested in studies of the topography and hydrology of the downtown area — architectural design firms flourished, highway construction became an advanced art. On the other hand, in line with our earlier discussion of the commoditization of research, almost no money or attention was spent in evaluating the life of the neighborhoods or how people might be affected by the abrupt changes. Only after demolition and reconstruction had already occurred and many viable neighborhoods had been cut in half by highways or simply demolished did social scientists, working with much less support than the engineers and economists who designed the projects, study the impacts of urban renewal. They spoke with survivors and recounted and documented what their lives were like before and after their removal.[5]

What happened during this period of urban development in the U.S. happens continually in more or less dramatic ways wherever commoditization affects people's lives by undervaluing and marginalizing the less-commoditizable goods and services that form the substance and glue of community. A small community, urban or rural, builds up a certain degree of self-sufficiency in the amenities of daily life. Many human needs can be met by neighborhood or local merchants. Extended families and tight-knit communities create networks of sharing, passing on of hand-me-downs, and other modes of recycling and conservation. Without such community and family ties, what was once shared and given away is now purchased. People need cars to get to the store or to visit friends or, lacking good means of transportation, they simply have nowhere to go. Even in the U.S. a third of the people do not or cannot drive on account of age or youth, disability, or poverty. Forcing everyone into cars disenfranchises many people from the full privileges of citizenship, and is psychically alienating too. In general, even as per capita productivity and consumption rise, many lives are contracted and more impoverished. Their contributions to the gross national product notwithstanding, the felt value of individual lives can diminish in proportion to the extent to which forces of commoditization govern their life chances and choices.

By oppression I mean the systematic denial of certain groups of peoples' development opportunities, while systematically directing material, energy, and attention preferentially toward privileged goods and people. The choice made by a society to develop a massive state-supported infrastructure to support the use of the personal automobile while simultaneously underdeveloping public transportation is an example. The systematic privileging of commodities and underprivileging of common goods results in those qualities and that "know-how" that are not readily commoditizable being systematically starved for material, energy, and attention. But these very qualities and skills — local connection, conviviality, ecological attentiveness, cultural and individual distinctiveness, and long-term care and maintenance — are inherent human qualities, and their gradual impoverishment by the free reign of commoditization diminishes all human beings. In this way, everyone is oppressed by the effects of commoditization.

Oppressed groups are affected in two distinct ways: First, by receiving a disproportionately small share of the benefits of commoditized economic development and, second, by suffering the greatest losses resulting from the deliberate underdevelopment and destruction of common goods, traditions, and values. By analyzing the socially and economically distorting effects of commoditization, we can more clearly define and understand the reasons why certain groups — indigenous people, people of African descent, other people of color, women, young people, and working-class people — are each oppressed and disadvantaged is some way associated with the effects of commoditization. Commoditization can also explain why and how the experience of this oppression continues and is, ironically, intensified even while important advances are made in the spread of freedom, democracy, human rights, literacy, and appreciation of cultural diversity in much of the world. Lastly, empowered with an understanding of commoditization, the road to liberation becomes clearer if not necessarily easier to travel.

5.4   Conquest, money, and commoditization

The origins of oppression lie in the history of conquest and enslavement and in the commoditization of money. This section takes a short sidetrack to explore the connection between conquest, money, commoditization, and oppression.

In general, when people, individually or in societies, prosper, it is either by luck, hard work, or stealing the products of another's luck or hard work. It is revealing to consider how different societies obtain the resources they need to develop and how their development is determined by which of these paths predominate. Civilizations throughout history have followed five principal paths of development or modes of accumulation:[6] (1) harvesting local resources in more or less knowledgeable, efficient, and economically mature ways; (2) nomadic movement, exploiting the available resources in one area and then moving on to the next; (3) trading, by building surpluses in one resource to trade to others with surpluses of a different kind; (4) finance, providing money or other resources with the expectation of being paid back with interest; (5) plunder, conquest, and war, stealing the resources of others. Most civilizations have relied on all of these in one way or another.

Many of what we now call indigenous cultures developed the skills, knowledge, and social systems for obtaining the food and raw materials they need to survive from the natural resources of their territory. This type of society develops through long and careful observation and experimentation. It has deep cultural roots. It draws first from the usable resources immediately available for gathering or obtainable by hunting. Success requires intimate knowledge of the properties of wild foods and beginning knowledge of how to encourage the growth of favored and most nutritious plants. Knowledge of the growing cycles and the potential value of different parts of a plant at different times in its life cycle is essential, as is knowing which plants are toxic or deadly. The ways of the hunted animals must be observed and remembered, and all the information somehow stored and communicated across generations. When and if the population grows too large for the available resources, people must split off and move into different territory and learn new ways of living over time, or they must develop new skills for tapping the earth's bounty by deliberately seeding, tending, and manipulating growing conditions to favor certain crops over others: the origins of agriculture.

If hunting and gathering or settled agriculture is successful enough to produce a surplus over and above what is required just to survive, then it becomes possible to support people for tasks other than food production or basic household and community maintenance. Some people become involved with trade and commerce. The trade and exchange of goods and services has always occurred among and between members of one's community and with neighboring communities and distant traders, but this trade has usually been but a small portion of what provides for the necessities of life. Social systems mature and develop in much the same way as any system matures, through increasing efficiencies in exchanges between people and their environment. There is no such thing as commoditization to distort this development because markets play such a minor role in the overall economy and there is no systematic privileging of marketable goods. Throughout history until at least the middle of the nineteenth century, the exchange of goods and services occurred within the context and bounds of a wide range of social rules and constraints. Decisions about the allocation of a community's resources and whatever surpluses it may have generated were determined by other than market considerations.

In general, nomadic cultures develop in ways similar to settled cultures, but in harsher environmental circumstances that make settled, locally self-sufficient life impossible. These nomadic societies become adept at conservation and utilization of scarce resources and skilled and efficient in the technologies of transport. They develop a broad and detailed knowledge of the territories over which they roam and the seasonal variations in resource availability over a large territory. Other wandering peoples may be people who chose to leave or were forced out of overpopulated or depleted home territories. Nomadic and other wandering cultures frequently evolve into trading cultures as they venture farther afield and encounter the homelands of diverse peoples. Information networks are established and people learn about the inventions and innovations of distant cultures. Development accelerates through exchange of information. Innovations spread and are adapted to new circumstances. Trade grows as people learn that natural resources and industrial skills are unevenly distributed throughout the trading region, and what is scarce here may be abundant elsewhere, and vice versa. The raw materials for basket-making is more abundant in one region while the shells used to adorn clothing is readily available from a distant neighbor. The shell-rich and basket-poor people are better off collecting more shells, which are abundant and easy to obtain, than trying to grow sweetgrass in their saltwater estuary. The basket-rich people need not settle for the bland and rare freshwater mussel shells when they can trade their abundant baskets for multicolored shells. In this way, trade makes both parties better off, as each obtains more easily what would be difficult to produce or obtain themselves. Development through trade is an expansion of the model of development through increasing efficiency, efficiency meaning the most satisfaction per unit of material and energy expended. This type of trade has a long history among peoples worldwide and did not begin, despite the Eurocentric myths to the contrary, with the era of European exploration and colonization of Asia, Africa, and the Americas.

Intensive localized settlement and nomadic wandering both depend on deep, intimate knowledge of natural cycles and nature's productivity. Both depend for their survival on not overexploiting the resources on which they depend. The trader relies less on knowledge of the natural world than he does on human nature. The success of his enterprise and his long-term survivability depend on his ability to gain the cooperation and trust of trading partners and on establishing and maintaining these long-term relationships. In other words, the development of each depends not only on the products and services that can be exchanged, but also on those "goods" and community and interpersonal services which cannot be bought and sold but which enhance the climate for fair exchange. Each group must invest at least as much time, attention, and resources on noncommodities as they do on what we would consider commodities. Just as importantly, in these traditional trading systems, the commodities themselves typically embody only a fraction of the energy and raw materials used to produce commodities in an industrial economy.

The interaction among cultures has gone on throughout human history. There has been a constant flow of knowledge and culture as well as goods and services. The relationships between cultures can be more or less symbiotic, in which each party enhances the other's capacity to thrive, or more or less parasitic, in which one party prospers at the expense of the other. Most intercultural relationships have been some of each. Trading and conquest have often gone together where periods of mutual interdependence are broken by periods of war.

Conquest has been the other, and perhaps most successful, path to modern development. The success of conquest lays the groundwork for the success of commoditization. For the conqueror, knowledge of the natural productivity and cycles is less important than knowledge of transport routes and the tools and technology of war. What becomes most important is the ability to transport large numbers of soldiers and their equipment along with the loot they are able to take. If anything, local ties and family loyalties become a hindrance to organizing a conquering army always on the move. Most important is the ability to stake and defend a claim to the products of others and to distribute the rewards of conquest among allies and supporters over long distances. More important than production for local consumption and self-reliance is production of the tools and weapons of war. People engaged in conquest have little time, attention, or need to produce what they can obtain from others. The rewards of production shift from the producer to the conqueror. Work and productivity yield to plunder.

Organizationally, above all else, the economy of conquest requires a currency whose value could and would be backed up by the power of the conquerors. Conquest and imperialism hastened the commoditization of money, while commoditization enhanced the power and mobility of conquerors. Commodities do not exist as commodities (things that can be bought and sold) until money exists. Commoditization as a selection force and allocation principle does not begin to distort development until money or currency begins to be disassociated from actual material relationships; in other words, the first step in the long road we have traced to the modern effects of commoditization is the commoditization of money. The earliest forms of money were objects of trade themselves, usually the least perishable of valued goods such as shells, precious metal, pelts. The value of other goods came to be defined and denominated in terms of the number of shells, weight of metal, volume of pelts, etc. Over time through the evolution of trade, various forms of money were selected out in favor of precious metals, whose relative rarity, malleability, and actual use in luxury items made them both desired, portable, and of generally recognized value. The commoditization of money follows the same path as all other examples of commoditization: money became more abstract, more portable, and less connected to any particular place or set of relationships. First money was the actual precious metal. Then it transformed into notes and bills, which represented claims to a particular store of precious metals being held to back up the value of the paper money. In the early twentieth century, the connection with actual stores of metals was lost, and instead the value of the currency was based on trust in the stability and power of the issuing nation's authority. In the late 1940s, all currencies of governments participating in the Bretton Woods institutions came to be denominated in U.S. dollars, but this system too collapsed in 1974, when currencies were allowed to "float" in value in comparison with other currencies in a free market exchange. In this process, what we have come to call the economy has become exclusively what can be exchanged in some way for currencies or other instruments representing abstract value. The meaning of efficiency in human activity shifted. No longer did it mean getting the most use and service out the least expense of material, energy, and time but instead efficiency came to mean the greatest output or yield per unit of value or money expended. As economic forces come to dominate more and more aspects of our lives (the commoditization of everyday life), access to cash becomes the sole source of capacity to provide the means of survival. What cannot be readily exchanged for cash becomes essentially worthless, and thus through commoditization comes the marginalization and impoverishment of noncommodities.

The commoditization of exchange values, or money, represents the capacity to centralize power and control resources in a more efficient way. Commoditization is made possible by currency becoming less associated with specific material exchanges and more a representation of power and authority. Although neo-liberalism provides a justifying ideology of free markets that gives commoditization the aura of a virtue, the political beneficiaries of commoditization are the powerful class in any society, whether capitalist or not. Alienating local resources and environments from diverse local peoples and measuring all activities according to their contribution to national product greatly facilitates the subjugation of conquered peoples and centralization of state power. This is why the major attempts at state social- ism in the former Soviet Union and Eastern Europe were initially very successful at stimulating economic growth and also devastating in the way they undermined rural and indigenous cultures.

5.5   Commoditization and the oppression of indigenous people

For hunters and gatherers, settled cultivators and nomadic people, life abso- lutely depends on an intimate, detailed knowledge of their particular envi- ronment. For such societies to continue as highly developed cultures, they must have the capacity to communicate and educate their young into this knowledge. This is why the environmental change wrought by industrial- ization and the social change wrought by commoditization is so damaging to native cultures and societies. These changes modify local, regional, and now global ecological patterns so as to make them largely unrecognizable, except in patches of undisturbed areas. This has the effect of short-circuiting the continued development of these cultures and represents both physical hazard and psychological devastation for native peoples. It is not that native people stop developing in the face of dramatic changes in their way of life: in the face of enormous challenges, native communities and people continue to apply their particular mode of understanding the world to the new con- ditions of industrialization and commoditization. But a tremendous amount of personal and communal energy must be expended in defense of their way of life against an oppressing society that does not value the noncommod- itizable aspects of community and spiritual life of the native peoples.

Highly evolved indigenous cultures are neither underdeveloped nor necessarily poor. The poverty of rural and indigenous people, although too often abject and devastating, is also mythologized as a direct result of the confusions of commoditization. Since only that which can be bought and sold is counted in the statistics that measure poverty, people who consume much of what they grow rather than purchase processed foods at the store, who wear clothing made from homespun materials, who educate each other's children rather than paying teachers, who build houses and barns with the freely exchanged help of their neighbors, etc., are regarded as poor, regardless of their actual condition. There is no distinction made in the poverty statistics between such successful self-sufficiency and the real dev- astation of people who are dependent on but living at the margins of market economies. Such people have precious little cash or things to sell with which to obtain the necessities of life. As commoditization spreads and self-suffi- cient cultures and communities are destroyed, more people move from autonomy to real poverty. In this way, even as per capita gross domestic product increases throughout the world economy, poverty grows, deepens, and becomes more miserable.

The situation for indigenous peoples (any group whose ways of life have coevolved over extensive periods of time in a particular ecosystem) is directly related to the effects of commoditization which selects out an important component of their very being, their connection with the ecosystem within which they have coevolved. In addition, inasmuch as the resources of their lands have been degraded by participation in the commodity economy, their previous way of life is often impossible to sustain as it once was. Since the commoditized economy has expanded to the point where its allocation structures totally dominate the distribution of materials, energy, and human attention, there is almost no chance to opt out in favor of traditional ways. The forms of knowledge valued by indigenous people — knowledge of relationships between and among the varied components of the home ecosystem — is largely unvalued by the commoditized economy for the reasons we have discussed above. The end result is a form of cultural genocide in which groups of people are systematically impoverished.

There is no doubt that tremendous potential exists for sustainable economic and cultural development in indigenous communities, but the realization of this potential will require substantial investments in research and development directed by indigenous peoples themselves. But these investments are systematically outcompeted by commodity-driven investments. Native American communities have been left not only with degraded ecosystems and the effects of cultural imperialism, but they have also been given few opportunities in the dominant, commoditized economy. As a result, many tribes have now chosen to exploit their limited sovereignty to introduce gambling onto reservations and to specialize in the sale of cigarettes and gasoline, which are untaxed on tribal lands. Focusing on these quintessential commodities (gambling, a commoditized response to hope; gasoline, a commoditized form of energy; and nicotine, like other drugs, a commoditized form of longing) has come at a high cost to the tribes themselves: in exchange for money that they now need to survive, they have become even more dependent on the commoditized economy and have admitted morally dubious and corrupting enterprises onto their territories and into their midst. The long-term effects of this phenomenon have yet to be seen, but it may make the tribes even more vulnerable to the homogenizing pressures of the dominant society.

As economic growth increasingly transforms the Earth, it further disables the capacity for traditional peoples to thrive, because their prosperity intimately depends on their knowledge of the productive processes of local ecosystems. As these ecosystems become altered by climate and other changes, indigenous knowledge becomes less relevant. There are countless examples. For one, consider the Bontoc people of a mountainous region of the Phillipines. They have developed a highly productive system of rice production in terraced paddies on the mountain sides. Planting is timed to coincide with key environmental signals, among them the arrival of a migratory bird known as the *kiling*. If climate change or declining populations due to loss of habitat in the birds' wintering grounds or other environmental

problems should disrupt the annual arrival of the *kiling,* then key components of the Bontoc's knowledge base becomes useless or even destructive
to their ability to produce rice. This occurs even if there are no environmental
problems in their immediate surroundings. Consider the fate of the Inuit
and other Arctic peoples. The Inuit rely on subsistence hunting on the Arctic
ice to provide most of their protein. The organization of the hunt, training
of the hunters, and preparation of the slain animals are deeply embedded
in Inuit cultural rituals and social norms. The warming trend associated with
global climate change may reduce the size and stability of the ice on traditional hunting fields, and could force much of the game such as caribou
further north ahead of the forests advancing from the south. Under these
circumstances, through no fault of their own, the Inuit and Bontoc peoples'
traditional indigenous knowledge, their cultural and knowledge-based capital, grows increasingly irrelevant in a changing world. Multiply this by
millions of such changes and you can see how indigenous ways of life
become increasingly threatened, and with this, important aspects of human
knowledge and development are lost.

The lands and resources of indigenous people were colonized during
the period of expropriation and wealth accumulation known as colonialism and imperialism. The concept of commoditization gives us a new
understanding of this period. The explorers, conquerors, and settlers in
effect robbed the colonies of their natural resources in a prototypical
process of commoditization. Whatever goods could be transported to
world markets most easily were plundered. In the process, the relationships that had evolved over centuries between the native peoples and
their environment were left in tatters. Those aspects of the native way of
life that hindered commoditization in any way were regarded as backward
and uncivilized, and were suppressed or destroyed. We will examine this
in more detail below.

"Development" is not the same as commoditization and industrialization, despite the many myths to the contrary. Every culture and society
matures along its own independent course of development in the sense that
it is continually learning, continually experimenting and adapting to new
circumstances. Languages evolve, tools become more useful and efficient,
stories change, people grow, at least until some crisis overwhelms a people's
developmental capacity. In many cases throughout history societies have
collapsed, and many have disappeared as a result of foreign conquest, civil
wars or environmental disasters such as the exhaustion of soils or depletion
of other resources, local climate change, prolonged droughts, crushing
storms, and so on. Every society and civilization that collapses represents
the abrupt ending of a valuable experiment in how to organize and sustain
human communities in a particular environment. It represents an enormous
loss in knowledge and potential. We can only begin to guess at the innovations and creative contributions these societies could have contributed to the
wealth of knowledge and skills with which we face the challenges of the
future. The possibilities lost with every failed social experiment is the reason

why the belief in continuous human progress, whether of the capitalist or Marxist variety, is so inadequate for understanding history.

The unnatural disaster that struck much of the world beginning in the late fifteenth century was European conquest and global imperialism. The conquest of the Americas, Africa, and Asia resulted in, among many things, a global eruption of commoditization as the conquerors removed the resources that were transportable and marketable, such as precious metals, slaves, timber, seeds, furs, sugar, spices, and artwork. Reading accounts of the records European explorers kept, one is struck by how little they actually recorded beyond the goods whose market potential they clearly understood.

They ignored or suppressed what was culturally and locally specific such as the thousands of plants and animal products the native people utilized for their subsistence, horticultural organization and skills, language, knowledge and understanding of local environments. The indigenous people fruitfully utilized far more of their immediate environment than did the European conquerors, including many more species as food stuffs, sources for cloth making and construction. What the explorers and conquerors discovered were highly evolved civilizations with unique, successful, and locally appropriate cultures. The people were misperceived as savages and represented as such through the popular media of the time. Once the conquerors forcibly gained control of area and resources, they successfully directed the vast majority of their investments of time, attention, and capital to maximizing production of key commodities for export to the imperial centers in Europe.[7]

The gold removed from the Americas amounted to ten times the gold production of the rest of the world and immediately transformed the European economy, whose growth had been limited by the amount of gold in circulation. Between 1500 and 1650, the gold of the Americas added at least 180 to 200 tons to the European treasuries.[8] The gold glut affected the African gold traders, who had been the leading suppliers of European gold. These traders turned instead to the newly lucrative slave trade. The slave trade killed several million people and destroyed native societies throughout Africa.

In Spanish and Portuguese America in particular, gold for slaves became the exchange. Both the plantations in the lowlands and the mines in the highlands required large amounts of human labor. By the time economic development reached the point at which regular exchange between the Americas and Europe occurred, most of the lowlands were depopulated of Indians by Spanish violence or disease. Indians taken from the mountains to replace them proved to be particularly susceptible to the lowland diseases such as malaria and yellow fever. African slaves soon became the primary source of labor except in the newly discovered gold and silver mines in the Andean mountains. There the scant air proved too much for the Africans, and most died soon after arrival, and so the highland peoples who were acclimated to mountain life were impressed for mine duty. In the first years of the mines, 80% of the forced laborers died.[9] Although conditions varied between the English, French, and Spanish throughout the Americas, the general pattern of relationship between the conquerors and the native peoples were similar. Even after settlement, the Americas were largely treated as a treasure trove to plunder rather than a place in that to learn how to live.[10] This is a significant psychological distinction which marks the conjunction of conquest and commoditization, an attitudinal stance conducive to a highly distorted form of development, an environmentally unsustainable form.

In *Changes in the Land*, author and historian William Cronon was one of the first to document the differences between the cultures of the European settlers and that of the natives in the part of North America that came to be known as New England. In New England both Indians and colonists attempted to impose a kind of order on nature. The order colonists imposed had a great deal to do with seeing natural resources as marketable commodities, of ignoring or misunderstanding nature's potential abundance by seeing in nature only that which could be commoditized. This, in effect, transformed the environment from a set of ecological relationships determined by natural and random forces into a set of economic relationships with wholly different, and human-dominated, forces.

The Indians of what became New England sustained abundance through ways of life that reflected the diversity and cycles of nature. They manipulated the land and harvested on a broad scale. They had highly effective techniques of fire management, population control, and portable settlements that could move with the cycles of land productivity. Indian technology and social divisions of labor, because they were finely adapted to natural systems, were far more subtle than what the colonists brought. So subtle, in fact, as to be invisible to the colonists who judged Indian ways from their vantage point. The fact that the Indian ways of life were so efficient as to leave far more time for leisure than the Europeans had ever known was mistaken for and mythologized as Indian savagery and sloth.

Because for the colonists of New England settlement was fixed and for Indians it was mobile, they each had very different notions of property. Fixed settlement leads to placing boundaries of ownership on the land, leading to the commoditization of land itself. Different understandings of property led to misunderstandings about property rights and inevitably to conflict. These processes led to gradual expropriation of Indian land and their impoverishment. Cronon sums up his argument, "Indians could not live as Indians had lived unless the land was owned as Indians had owned it."[11] Which is to say that Indian ways of life became impossible under conditions of colonialism.

There was a tremendous disparity between the abundance of natural resources — particularly timber and furs — in America and the paucity of the same in Europe. The contrast between New England abundance and Old England scarcity created a kind of energy differential, which set up a current of commodity flow that rapidly depleted the colony. Because the money value of resources were made artificially high by being priced according to Old England realities, the return for labor was high. There was little incentive to invest time and capital in husbandry. Looting was easier and more reward-ing than careful cultivation.

Even under these conditions, and suffering from the new European-bred diseases, which in some cases wiped out as much as 95% of the population, indigenous people continued to adapt to their new conditions and develop long after the European conquest. The example of the inde-pendent and resilient Cherokees is not atypical. Generations prior to the coming of the Europeans, the Cherokees had developed an advanced society around settled agriculture and elaborate social institutions. With the coming of the Europeans, the Cherokees managed to adjust to their presence while successfully preserving their own culture and institutions. They invented a way of writing down their own language based on written English. The U.S. Supreme Court had acknowledged Cherokee sovereignty over their own lands, and recognized their "civilized status." But gold was discovered in Cherokee country and in 1838 the Cherokees were pushed off their Carolina homelands and onto a forced march to Oklahoma, where they were given a reservation. So many perished along the way that the journey became memorialized in Cherokee and U.S. history as the Trail of Tears. Once in their new lands they began to recreate their communal life. By the 1880s their independent economic and social development was well under way on the reservation. But then they were forced by the 1887 Dawes Act to abandon their traditional institutions of tribal land ownership in common and transform their new territory into a collection of commodities by divid-ing the land into parcels and assigning property rights to individual owners, effectively ending tribal authority to manage their communal destiny. What becomes obvious here is that outright oppression through force, coercion, and extirpation go hand in hand with the structures of commoditization that we normally think of as more benign, such as law and private owner-ship. In this case, the law was used to destroy a people for the benefit of economic growth and cultural dominance.

Even the notion that international trade and development was a Euro-pean innovation is misleading at best. The year 1498, when Vasco da Gama arrived by sea in India and opened the coveted sea route to the Indies, is frequently heralded in European textbooks as the welcome beginning of world trade and sometimes as the discovery of India. But long before da Gama landed in Goa, there had been considerable contact and trade between the principalities of India and Europeans across the overland route, a route controlled by Muslim powers and merchants. India was already a meeting place of civilizations and cultures and goods routinely moved across India, Africa, and the far East. Already in parts of India, a huge shipbuilding industry existed. The Portuguese conquest and colonization, followed later by the English, deliberately set out to systematically de-develop the conti-nent, barring trade between the native people and anyone other than the Europeans, taking control of local resources and beginning a determined effort to substitute Europe's way of doing things for native life. Speaking about the results of European imperialism and commemorating the arrival of da Gama, S. M. Mohamed Idris, President of the Third World Network, concluded that:

> While European economies and societies benefited out of all proportion, the cultures and civilizations of Asia, Africa and America were either physically exterminated or cast as inferior contributions to human history and in need of total replacement with European institutions and languages.

> This, in fact, is the fundamental difference, in India at least, between all the invasions prior to 1500 and the invasions inaugurated by Vasco da Gama. The earlier invasions enriched India and India enriched the invaders. The European invasions enriched only Europe, impoverishing India instead. The enrichment of Europe was in terms of stolen and violently plundered wealth and hence superficial. As a consequence of the disruption of the societies of Asia and Africa, human history — including European history—was equally impoverished. I am saying this because we are now convinced that acts of oppression always impoverish not just the oppressed but also the oppressor.[12]

5.6   The legacy of colonialism in the modern global economy

This mutual impoverishment, this stunting of development potential, remains the terrible legacy of underdevelopment brought about by colonialism and commoditization. Everyone loses from the stunting of diverse optional pathways to human and economic development. The great and growing disparity in the distribution of wealth and control of resources among people and cultures in the world begs some kind of explanation. One such explanation rests on the ideology of racism, which purports to identify inherent differences in work attitude and ambition among the races.[13] A second explains existing inequalities as the result of the history of imperialism and conquest, which destroyed viable civilizations throughout the world and left a lingering legacy of poverty and underdevelopment that has yet to be addressed.[14] A third focuses on the effects of geography, environmental factors, and historical contingency to explain why one culture ultimately predominates over another.[15] The concept of commoditization can help reconcile these different approaches to history. Yes, chance and geography play a major role in history, just as chance and geography play a major role in all evolution, but once financial and political power become merged, commoditization becomes an active selection process steering the allocation of a society's energies toward specific forms of exploitation and conquest. The ideologies of racism were based not only on distrust and fear of difference, but also on real differences related to the effects of commoditization. Cultures that have evolved under the influence of commoditization have overdeveloped certain traits associated with acquisitiveness, competition and restlessness — ills, interpreted as virtues, that less commoditized cultures have successfully avoided. The traits encouraged by commoditization have nothing to do with the capacity for hard work, ambition or any other human quality, but rather they are associated with the overdevelopment of certain human traits that serve the ends of commoditization at the cost of the loss of many other valuable and important aspects of human nature that do not. The confusion at the base of almost all discussion of economic development is that it holds the commoditized version of overdevelopment as the norm, as the goal every society should achieve.

Much of the subject known as development economics[16] attempts to address the question of how economic development occurs or fails to occur. Almost all economists who study development agree that, as one economist put it, "economic development is not equivalent to the total development of society: it is only a part — one dimension — of general development."[17] But regardless of this understanding, development has come to mean the domination of economic values in the social and political life of a nation and the incorporation of national economies into the global market. This invariably unleashes the commoditization pressures we have discussed and leads to distortions in development typical of commoditization. These include the industrialization of agriculture, the emphasis on export of raw materials and consumer goods, investments in large, centralized facilities for electricity production, and the emphasis on capital-intensive and high-technology industries. In general, these require large inputs of borrowed foreign capital and produce demand for technically skilled labor, but produce far fewer jobs for those with fewer commoditizable skills. Two outcomes can be predicted: huge amount of foreign debt and vast inequalities in wages. The elite classes (often the lightest skinned and most Europeanized of the native populations) concentrate more of the power and wealth. These make up outposts of wealth in a vast sea of poverty.

The foreign debt creates the need to earn hard currency (usually U.S. dollars) in order to repay the debt, which stimulates ever-increasing commoditization in the home economy. All noneconomic investments, including environmental protection, become drains on the economy in the rush to increase exports. The concentration on exports usually means raw materials in the form of increased mining and logging, as well as increasingly low-cost consumer products. With much of the former colonial world following the same path of development at the same time, a glut of primary commodities and consumer goods hits the markets, driving down prices and necessitating even more production in order to earn the same amount of dollars. In the meantime, the economic prescription of choice promoted by the International Monetary Fund and the World Bank is total commoditization in the form of slashing public expenditures on everything but the commoditized economy, eliminating the subsidies for basic foodstuffs and public transportation, deregulating the market in the provision of public services, and privatizing much of the economy including that typically sheltered from market forces (such as public utilities and natural resource development). The crisis of development is bound to worsen as the self-reinforcing commoditization machine gains global momentum. The spread of commoditization is the spread of the exploitation and oppression of indigenous people, women, children, the elderly, and the disabled.

5.7   *Commoditization and the oppression of women*

Expectations in many, if not most, cultures is that women attend to the least commoditized spheres of life: childcare, day-to-day maintenance of home and relationships, care for sick and elderly, and the skills and investment of time and attention into interpersonal relationships — what could be called the *economy of care and connection*. Women's economic innovations and development are often directed to that which is most resistant to commoditization.[18] As the economy becomes more commoditized, the distribution of material, energy, and human attention becomes more dominated by commoditization pressures, and the least commoditizable spheres of life become comparatively underdeveloped. The economy of care and connection, which had been defined as the women's sphere and to which women had been limited by social expectations and rules, has itself become commoditized to the extent possible. Women have moved *en masse* from unpaid household work to salaried employment, largely in the area of women's paid work as health care workers, service providers, retail sales people, day care workers, and other low-paying tasks associated with the newly commoditized economy of care and relationship.

Women are saddled with unequal responsibilities for delivering essential community and family services. Their availability for service in the commercial economy comes at a high price. Working women tend to spend far more hours in the economy of care and connection even when they are fully employed as salaried workers.

At a time when the U.S. economy was less commoditized than at present, income was expected to be distributed so that a husband/worker could support both himself as participant in the sphere of commodity production but also the noncommodity spheres of his household and community, the so-called "family wage," which was meant to also support the less commoditized or women's sectors. As the parts of the economy with low commodity potential have become increasingly marginalized, this is no longer viable. Hence, the sphere of "women's work" has grown increasingly impoverished to the point that women and men can no longer afford to spend much of their time there unless that work too is commoditized and becomes paid labor in the childcare, health care, domestic maintenance, counseling, landscaping, and other sectors of the economy of care and connection. As we have seen, in each of those sectors the most commoditizable components of the work — facilities, play equipment, cleaning supplies and tools, pharmaceutical drugs, etc., receive the greatest amount of investment. The trend through increasing commoditization is for the economy of care and connection to first become part of the public service sector and then to be fully commoditized and privatized as part of the commercial economy.

This phenomenon of shrinking noneconomic space accompanies commoditization and exacerbates the difficulties of those who by choice or necessity remain in the non-commoditizable sectors. It complicates the politics of women's liberation by segmenting it into at least two camps with conflicting political strategies: those who want to make more room for women within the commoditized economy and those who resist the effects of commoditization and demand that more of society's resources be directed to the economy of care and connection (see Chapter 8 for the differences this makes in policy prescriptions). Freeing women to participate and gain in all aspects of economic life is important but liberation from the oppression associated with commoditization itself will only be possible when the economic forces of commoditization are addressed directly by redirecting greater portions of society's resources toward the economy of caring and rewarding those, both women and men, who freely choose to use their time and attention toward the noncommoditizable activities of care.

5.8   Commoditization and class oppression

As the economic space available for independence and economic self sufficiency grows smaller, the numbers of people who must rely mostly on the sale of their labor, time, and attention also grows larger. Large numbers of workers, however, are also investors and derive income both as owners and laborers. They benefit from the continuing economic growth associated with commoditization. This solidifies a large group with middle class buying power that benefits from returns on investment. It also creates markets for services that enhance the lives of this middle class and, in part, substitute for the losses of community and extended family and a solid connection to a particular place on Earth. The politicization of class identification has grown more difficult as a result. The few remaining industrial workers earn salaries typical of the middle class, while many professional and white collar jobs carry comparatively low wages.

Commoditization always favors automation over the improvement and development of workers skills in relation to the task at hand. Machines invariably carry more of the qualities of a commodity then do workers and are therefore privileged goods when it comes to research and development. The first jobs to be automated or transformed through commoditization are the manual and menial tasks. But with the commoditization of organization and management, which has largely driven the evolution of computer technology, many of the tasks associated with the middle class have also been taken over by machines. The result is increased labor productivity (economic output per worker hour) and a growing economy. The wealth produced by this economic growth is invested in further commoditization, largely in the service sectors which tend to be more difficult and less suited to automation.

Because commoditization is the privileging of commercial goods and services over all other aspects of life, when commoditization determines most of the investment in a society then it is economic growth through increased labor productivity to which society gives highest priority. Efficiency gains are measured in improvements in output per unit of labor or capital. As long as energy remains cheap, this always encourages automation and standardization over job creation and improvements in job satisfaction. All this adds up to the inexorable march of commoditization: jobs per unit of capital invested continues to decline while the strains on the remaining workers increase.

As the tools of production become more mobile and standardized (following the path of commoditization) and financial capital becomes more mobile and free from constraints of national policies (also following the path of commoditization), then capital is much more free, mobile, and portable than workers can ever be (especially if they are in any way tied to the noncommoditizable values of family and home). The only workers able to take advantage of this situation are the ones who are freest to move toward the factories built by these free flows of investment. In much of the world, it is the young men, who are flocking to the mines and massive industrial zones of the Third World, often leaving rural villages to be largely populated by impoverished women and children. In the wealthier parts of the world, the mobility of whole families is more feasible, but the resulting flows of population accelerates the homogenization of cultures and the breakdown of family and community ties, which then serves the commoditization of human relationships.

5.9   *The underdevelopment of imagination*

The most pervasive aspect of commoditization and the most stunting in terms of human development is the underdevelopment of the imagination and creativity that results. What might we accomplish if we put a good portion of our intelligence toward designing better and better ways of doing more with less production and getting more satisfaction out of less and less consumption of energy and materials, developing agriculture that works with natural cycles rather than against them, exploring the possibilities of alternative energy, discovering and promoting health in all its manifestations, and teaching and training our young people in the skills and sensitivities required to build a sustainable society? Environmental education and skills education for community building have made inroads in the schools as a result of parents' and teachers' growing appreciation for the kinds of skills and knowledge young people may need to contribute fully to finding solutions to the major social problems of our times. These skills have to do with creative problem-solving, listening and communicating, ability to derive fresh responses to each new situation, careful observation and aware-ness of the natural world, and many others. Larger social forces, however, stimulated and supported by the economic pressures of commoditization, operate in the exact opposite direction of these skills. Children quickly under-stand the workings of the most powerful economic forces in their lives. If all they have is their labor to sell, they must fit into some productive activity, which under the pressures of commoditization, must necessarily be the production of goods with high commodity potential. The only reason to develop other skills, be they athletic, artistic, or intellectual, is if they take those skills, usually to the exclusion of all others, to such a point of refine-ment that they can become professionals, celebrities perhaps and, in effect, commoditize themselves. Otherwise those skills must be reserved for leisure time which, again under the pressures of commoditization, gradually shrinks. What is lost as a practical idea is that everyone should receive a civic education to prepare to participate in a society in which they are expected to have the understanding and capacity to create beauty, recognize and protect ecoosystems, assist others in developing their own skills and intelligence, nurture diversity, and otherwise build and maintain community.

In a commoditized economy only the well-to-do have the privilege of utilizing their skills in the noncommoditized sectors of the economy, as leisure time or volunteer activity. Thus it is not surprising that the core skills of building a sustainable society are routinely offered in middle class schools, whereas the poor, in a quite rational response to the conditions created by commoditization, clamor for the basic skills needed to obtain a job. The pressures of academic competition appear to be particularly stressed by parents and administrators in low-income schools.[19] As a result, much like in the economy itself, the skills that are most difficult to transform into commodities receive far less of the rewards and attention of the education system. They become, like handmade goods, organic food, and alternative energy, luxury specialty items for the rich. The systematic distortion of education and the narrowing of the capacity to serve community in other than commodity production, is a form of oppression under which we all presently suffer. Young people are inherent innovators and naturally ideal-istic and generous. They also maintain high ideals and long for participation in a society that expounds high ideals and attempts to live up to them. Commoditization only rewards innovation that can be sold by the innovator and her backers for profit. This narrows the scope of potential innovations. It hits young people particularly hard as they consider their options for fully developing their human potential.

It is common to blame technology for these distortions of human devel-opment, but technology and the technological imagination in particular are as much a victim of this distortion as they are its perpetrator. There are all sorts of low commodity potential sectors that can benefit from improvements in gadgetry and design. Mass transportation, solar and wind energy, small-scale rural electrification, energy conservation, materials conservation and recycling, new methods for protecting wetlands, conserving water.... The list is endless. It is the oppression of technologists which forces them to use their skills and energies in only one direction, toward increasing the amount of material and energy moving through the economy which, given the trends in the environment, is the exact opposite of what needs to be done to protect and restore the ecological integrity of the earth. If people with technical skills and interests care about the earth, its people, and its creatures, which we all do, then the limited opportunity to put their skills to the service of what matters to them most is a significant form of oppression. Of course this is equally true for all of us who must market our skills and time in exchange for wages.

Human intelligence has the capacity to derive a multitude of potential solutions to the problems of providing for human welfare. Yet only certain narrow types of solutions, commoditized solutions, can predictably depend on the support of the larger human community in the form of investments of material, energy and attention. As a result they are the only ones to be considered "practical" economically or politically. This means that the vast potential of human intelligence is systematically stunted. This amounts to a form of oppression that damages all human beings, even though some classes and groups materially benefit from the economic growth that commoditization drives. These groups benefit largely because of historical circumstances that have resulted in members of their circle of relationships having greater access to the wealth produced by the current economic system and greater influence over the allocation of resources for research and development.

Whenever the skills of slow careful observation of the natural world, creative problem-solving, community-building, and peace-making are ignored or marginalized in the classroom to make more time and resources for improving performance in standardized tests, then both children and teachers are oppressed. Whenever childcare workers are paid less than toy salesmen, the children and the workers are oppressed. Whenever organic farmers earn less than pesticide producers, we are all oppressed. Whenever pollution or expropriation makes it impossible for indigenous people to live off the renewable resources of their land and waters, they are oppressed. Think of any important, constructive, and life-giving activity that is difficult to commoditize and you will discover that the people who provide those services are systematically denied the resources they need to develop their skills, innovate their methods, and thrive. The result is a pervasive form of one-sided maldevelopment.

It's not that people are not making enormous personal sacrifices everyday to hold communities together, to get to know and love particular places on earth, to provide personal support, comfort and care, to recycle and conserve despite the economic disincentives. The point is that these skills and services should and could be integrated fully into social life and the people providing them appropriately rewarded. In later chapters we will consider the public policies needed to do that and the social movements needed to create the political possibility for their happening.

Economies that give free reign to commoditization have been and likely will continue to be enormously successful at producing goods (commoditized ones). But this success has two major drawbacks: first, as long as commoditization is unrecognized and uncontrolled, development will be distorted in the ways I have described here, and second, to the extent that it continues to be successful it will continue to increase the rate of mobilization of energy and materials and all the environmental calamities associated with it and therefore cannot be sustained. One example: in 1995 there were slightly less than two people per car in the U.S.; in China and India, with most of the world's population between them, there are more than 270 people per car.[20] Given the density of population it is unlikely that these countries would develop in as totally auto-dependent a way as the U.S., but even the population-dense Netherlands, with three people per car, has 90 times more cars per capita then India and China. The combined population of Africa, Asia, Oceania, and Central and South America is 4.4 billion people. Among them they own 200 million motor vehicles (cars and trucks combined). In 1995 there were 200 million people in the U.S. and a total of 200 million motor vehicles.[21] If the heavily populated, underdeveloped countries begin to approach these levels of automobile dependence, than a 20-fold increase in automobile CO_2 emissions, ground-level ozone, carbon monoxide, and other air pollutants can be expected. Similarly, the per capita energy use in North America and Europe is well over 200 gigajoules; in India and China per capita energy use is around 20. The U.S., with 4.7% of the world population, consumes 25% of the fossil energy used worldwide. U.S. food production, processing, and distribution alone is capable of consuming all petroleum reserves in a handful of years were it to be globally replicated.[22] The average child in the Third World consumes, depending on what is measured, anywhere from 1/20 to 1/300 as much in the way of energy and materials as an American child.[23] If the rest of the world follows the industrialized countries in the path of commoditization, the increase in energy and materials use and associated pollution would be socially and environmentally devastating.

With commoditization limiting the possibilities for less consumptive development paths, it is hard to imagine anything other than economic or environmental collapse, or both, as a result of the Third World's aspiration to replicate the commoditized development path of the First World.

This brings us to the need for fresh thinking about development, the sort of fresh thinking that the concept of sustainable development was meant to be when it was first promoted. Sustainable development could be a renaissance of local economies, locally appropriate practices, tools, and social institutions that provide opportunities for material well-being and physical, cultural, and spiritual health for all people while at the same time protecting and enhancing the ecological life support systems and resource base upon which all health depends. This can be accomplished only by the institution of deliberate social and economic policies that provide a countervailing development pressure to the effects of commoditization. These policies will be discussed in some detail in Chapter 8.

5.10 Conclusion

This chapter has described the relationship between commoditization and oppression. If this relationship is true then the same public policies and changes in priorities that are required to counteract commoditization will be effective in liberating people from the effects of oppression. We have already seen that commoditization is also a significant factor in global environmental destruction; it is possible that the same approaches needed to protect the environment are also the same policies needed to advance human liberation. This suggests an exciting potential for social movements built on an alliance between those working for the liberation of oppressed groups and those working on behalf of environmental protection and conservation. This theme will also be taken up more fully in Chapter 8.

In these first five chapters we have shown how commoditization works and some of the negative effects it has had on communities and the environment. Before beginning to consider public policies to counteract the effects of commoditization and social movements to bring those policies about, we will first look at the history of deliberate public policies that have encouraged commoditization.

Notes

1. Wolff, E.N., Recent Trends in Wealth Ownership, a paper for the conference on *Benefits and Mechanisms for Spreading Asset Ownership in the United States*, New York University, December 10–12, 1998.
2. How to Build Community (SCW, 1998) was published by Syracuse Cultural Workers, a national publisher/distributor of feminist, progressive, multicultural products. It is available as a 12" × 36" full color poster, T-shirt, notecard and postcard. To order or for a catalog ($1): SCW, Box 6367, Syracuse, NY, 13217; (315) 474-1132, fax (315) 475-1277. Visa/MC accepted.
3. Bellush, J. and Hausknecht, M. (Eds.), *Urban Renewal: People, Politics and Planning*, Doubleday, New York, 1967.
4. Fried, M., Grieving for a lost home, in *The Urban Condition*, Duhl, L.J. (Ed.), Basic Books, New York, 1963.
5. Fried (1963). See also Wood, E., *Social Planning. A Primer for Urbanists*, Pratt Institute, Brooklyn, NY, 1965.
6. Much of the ideas here are derived from the work of Jared Diamond, particularly *Guns, Germs and Steel*, W. W. Norton, New York, 1997, as well as the historical analysis of Karl Polanyi in *The Great Transformation*, Beacon Hill, Boston, 1957. The paths of development are similar to what is presented in Hornborg, A., Ecsosystems and world systems: accumulation as an ecological process, *Journal of World-Systems Research* 4(2), Fall 1998, 169–177.
7. Cronon, W., *Changes in the Land*, Hill and Wang, New York, 1990, and Sale, K., *The Conquest of Paradise*, Alfred A. Knopf, New York, 1990.
8. Weatherford, J., *Indian Givers*, Crown Publishers, New York, 1988, p. 10.
9. Weatherford (1988).
10. Cronon (1990).
11. Cronon (1990), p. 80.

12. Idris, S.M.M., The message from Calicut — 500 years after Vasco da Gama, *Third World Resurgence*, No. 96, August 1998, p. 21.

13. Europeans in colonial times typically described the "natives" of these countries as superstitious, lazy, careless, unenterprising and merely survival-minded. These derogatory views were part of the rationalization of the prerogatives they took to themselves and were able to retain as a supercaste." Myrdal, G., Asian Drama: An Inquiry into the Poverty of Nations, Twentieth Century Fund, New York, 1968, p. 62.

14. See, for example, Sale, K., *The Conquest of Paradise.*, Alfred A. Knopf, New York, 1990. Also, Editors of the Ecologist, *Whose Common Future? Reclaiming the Commons: the Ecologist,* New Society Publishers, Philadelphia, 1993.

15. Diamond (1997).

16. Ray, D., *Development Economics*, Princeton University Press, Princeton, 1998; Meier, G.M., *Leading Issues in Economic Development*, 5th ed., Oxford University Press, New York, 1989; Rostow, W.W., *The Stages of Economic Growth; A Non-Communist Manifesto*, University Press, Cambridge, 1971.

17. Meier, G.M., *Leading Issues in Economic Development*, 5th ed., Oxford University Press, New York, 1989, p. 5.

18. Waring, M., *If Women Counted*, Harper, San Francisco, 1998. See also Hynes, P.H., *Taking Population out of the Equation: Reformulating I=PAT*, Institute on Women and Technology, North Amherst, MA, 1993, and many others.

19. Kohn, A., *No Contest: The Case against Competition* (rev. ed.), Houghton Mifflin, New York, 1992.

20. American Automobile Manufacturers Association, *Motor Vehicle Facts & Figures 1997*, American Automobile Manufacturers Association, Washington, D.C., 1997, pp. 45–46, http://www/aama.com.

21. Quoted in "Matters of Scale: Driving Up CO_2" *WorldWatch*, November-December 1997.

22. Pimentel, D. and Pimentel, M. (Eds.), *Food, Energy and Society,* University Press of Colorado, Niwot, CO, 1996, p. 291.

23. Hall, C.A.S., The environmental impact of U.S. babies, *Earth Island Journal*, 10(4), 19, Fall 1995.

chapter six

The institutional development of the commoditized economy

Alex Antypas and Jack P. Manno

Contents

6.1 Introduction

The strength of commoditization pressures and whether countervailing forces exist are determined by the structure of the economy, the way contradictory economic forces are encouraged and/or restrained by the political, social, and legal institutions of a society. The modern industrial economy has evolved to the point in which market forces predominate over other social forces, granting enormous power to the forces of commoditization to affect our lives. In order to understand the implication of commoditization for environment and society we must first consider the historical conditions out of which the modern market-based economy emerged.

This chapter will describe some elements of early modern economic and political institutions in four European countries — the Netherlands, England, France, and Spain — in order to examine the emergence of the earliest and most basic institutions that either fostered or inhibited a growth oriented, commoditized economy and to show the relationship between the economic and political systems. The chapter will then go on to examine some of the major institutional changes that occurred both prior to and during the Industrial Revolution that made that revolution possible. Finally, the specific American experience will be considered as a case of political and social organization fundamentally oriented toward organizing the productive capacity of a population toward economic growth.

6.2   *European foundations*[1]

From the mid-fifteenth to the mid-seventeenth centuries, Europe underwent fundamental economic, political, social, and cultural transformations that laid the social and political groundwork for the modern economy. The spread of Renaissance humanist thinking expanded human possibilities in all spheres of life from the bonds of medieval orthodoxy. Humanism opened the imaginations and curiosity of artists, philosophers, and scientists who undertook efforts to improve the lot of humanity through intellectual and material changes that would have been resisted during the heyday of medieval society. The exploration and conquest of the New World, trade with the Far East, and the export of slaves from Africa caught the rest of the world in the economic forces of European expansion and ushered in a period that established the basic framework for the modern capitalist economy. All of these changes worked together and reinforced each other, and in time conjoined to undermine and overcome the structures of feudalism, resulting in the institutional framework of the modern global economy.

The path from feudalism to modernity was uneven, often circling back, and took different turns for different countries. Of the major states in Europe, England and the Netherlands developed political and economic institutions most conducive to breaking the political and economic strictures of feudalism and releasing the human energy required for continuing economic growth. Spain and France, on the other hand, adopted policies that inhibited growth in the long-term and kept those countries on the margins of the emerging modern economy for some time. In France this course shifted drastically in 1789. Spain continued to languish economically relative to other European states until recently.

6.2.1 France and Spain

When Charles VII became king in 1422, France was suffering from the aftermath of the devastating Hundred Years War. It was partially occupied by the English, bandits and roaming bands of soldiers terrorized the countryside, and the aristocratic houses were in a constant state of strife with one another. Charles faced the task of reuniting the country and reestablishing social stability and order. In order to do so he had to raise revenue from the populace. The Estates Generale, the representative body nominally responsible for taxation, granted the Crown temporary emergency powers to levy taxes. For all intents and purposes, these temporary powers were never rescinded, and the King became the sole tax-imposing authority.

Furthermore, the French king had the right to grant and change property rights, which, coupled with his authority to tax, gave the crown near exclusive control over the development of the nation's economy. In order to secure tax income, the king made it a practice, for instance, to trade property rights for taxes, which turned out to be a drag on economic growth for centuries. The crown found a ready partner in craft guilds which had established themselves throughout the towns of France. Guilds were organizations of craftsmen who banded together to guard entry into the profession and to minimize competition. Guild controlled crafts were able to keep prices high and to keep outsiders out. In return for steady taxes from the guilds, the crown guaranteed their right to control local markets, effectively creating state-protected monopolies. The tax system was administered by a large bureaucracy which could then be used to expand regulatory control over the economy.[2]

The bureaucratically enforced guild system allowed the French economy to remain regional rather than national, thereby passing by the efficiencies that might have been gained in a larger system. Moreover, the guild system depressed competition and innovation, and the tax bureaucracy controlled a significant portion of the profits that the economy produced. In the long-term the bureaucracy also became a powerful political actor in French politics, a role it continues to play to this day.

The economic experience in Spain leading up to the modern era was similar to that of France. As in France, the Spanish crown managed over time to centralize the authority to tax and grant property rights in itself. However, before Spain realized the economic costs of this system, it first went through a period of great, almost hegemonic power.

The colonies of Spain, extending from the Americas to the Low Countries and Milan and Naples, provided the Spanish crown and the Spanish economy with an influx of taxes and bounty, driving growth and an expansionist military policy. However, Spanish military expenditures easily matched growth in revenue. In other words, the costs of the empire threatened to exceed the economic benefits of maintaining it. When the Low Countries revolted, Spain lost its most important source of revenue. Coupled with declines in gold from the Americas, the effects on Spain were catastrophic. In response, the crown made similar arrangements with guilds in Spain as the French crown had made in France. The combined result of high military expenditures, the loss of the Low Countries, declines in confiscated wealth from overseas, and the establishment of state-protected monopolies at home was that Spain entered a period of economic decline relative to other parts of Europe.

6.2.2   The United Provinces and Britain

In contrast to France and Spain, the United Provinces, encompassing today's Holland, Belgium, and Luxembourg, experienced sustained economic growth throughout the sixteenth century and through to the present. The United Provinces managed to escape the economic pitfalls of its more authoritarian neighbors and masters through a combination of private property rights and a trading system that made it the hub of international trade.

Until 1576, the United Provinces were a part of the Spanish empire, and therefore subject to economic controls established by the Spanish Crown. However, the Crown did not impose the kinds of restrictions that were operative in the home countries on its possessions. Rather then establishing monopolies as it did with the Spanish guilds, the Crown allowed a relatively free market in property rights to emerge among the Dutch.

The legislative body, called the Estates-General, was given the authority to levy taxes. Because many of the representatives in the Estates-General were merchants or from merchant families, the legislature was very amenable to establishing rules of trade and private property protections that were conducive to trade and economic growth. The Crown was happy to go along with the Dutch arrangement so long as it received sufficient revenues. In fact, the Dutch became the principal contributors to Spanish coffers, resulting in discontent in the United Provinces, and eventually a revolt by the seven northern provinces.

The Dutch excelled at developing innovative institutions that increased economic efficiency and expanded markets and private property rights. They centralized trading in auction markets in which price updates were efficiently circulated, conducted business with standardized contracts, and created courts exclusively for commercial disputes. Moreover, the Dutch established capital markets that developed new forms of tradable capital products. In other words, the Dutch found ways to make capital an easily tradable commodity.

The Low Countries also experienced a revolution in agriculture in the sixteenth century. Dutch agriculture became more capital intensive, which allowed farmers to expend greater resources in draining wetlands, which once covered the region. Dutch agriculture became heavily dependent on fertilizer and mechanical inputs. Also, the Dutch chose to take advantage of the efficiencies of specialization and were the first in Europe to grow regional monocultures.

Before any other nation, then, the Low Countries created a model for the modern, heavily commoditized economy. Not surprisingly, the Low

Countries were the economic powerhouse of Europe. The property rights regime in the Low Countries was similar to present-day property institutions, allowing a free trade in goods with enforceable commercial contracts and open competition among buyers and sellers. Markets were allowed to evolve along the most economically efficient lines, and farmers as well as traditional businessmen could specialize in certain products because the expansive markets of Antwerp and Amsterdam provided a rich trade in all available products. And unlike Spain and France, the state did not impose onerous, economy-depressing taxes. Rather, all trade was taxed at a low rate to pay for necessary state expenditures and revenue for the Spanish Crown. When the Low Countries rebelled against its Spanish rulers, first in 1566, resulting in widespread destruction and oppression, and then again finally in 1576, the burden of supporting Spain was lifted, allowing for an even greater rate of economic growth in the region.

The English economy developed along lines similar to the Dutch, with the consequence that England became the second nation in Europe to enjoy long-term economic growth in a market-driven economy much like today's economies. Eventually England, due to its larger population and greater abundance of natural resources, overtook the Low Countries as the economic leader of Europe. The ascendancy of England had been in no way inevitable, however, as the English economy and political system resembled those of France and Spain more in the fourteenth and fifteenth centuries than that of the Low Countries.

The English suffered the consequences of the Hundred Years War along with the French, as well as the sectarian divisions that resulted in the War of the Roses. English political and social insecurities made that country prey to the same kinds of royalist policies that drove the French to trade economic and political liberalization for stability and centralization. Indeed, the English monarchs pursued policies reminiscent of those of the French king, with the crown attempting to gain control over the distribution of property rights and taxation. However, unlike the Estates Generale, the English Parliament never gave up the right to control the rate of taxation.

The conflict of interest between the Parliament and the monarchy reached its peak under the Stuarts. When James VI of Scotland (James I of England), the son of the beheaded Mary Stuart, inherited the crown from Queen Elizabeth, he attempted to expand the rights of the monarchy to reflect what he thought was the divine right of kings to rule without constraints imposed by Parliament, the clergy, or any other party. Parliament successfully resisted the King's incursions on its traditional rights, withholding from the King the right to decree new taxes. The clash between Parliament and the Stuarts reached its climax under Charles I, who eventually lost his head to Cromwell's revolutionary dictatorship. When the monarchy was restored in 1660, Charles II, guarding his head by avoiding antagonizing Parliament as his father had, allowed certain crucial economic–political reforms to proceed that ultimately circumscribed the power of the monarchy to regulate the economy.

The reforms enacted by Parliament benefited the land-owning class first and foremost, but they had effects that went beyond the landed aristocracy in the long term. The essence of the reforms was the abolition of financial obligations owed by landowners to the King for the right to hold tenure over land. Instead, land was turned into privately held property in the modern sense. In exchange, the landed aristocracy took upon itself the burden of taxes defined and set by Parliament. Parliament then paid the Crown as it saw fit, effectively making it the central political institution in Britain for the next century and a half.

The reforms of 1660 increasing the scope of Parliament's political authority were broadened dramatically during the "Glorious Revolution" of 1688. The son of Charles II, James II, having turned decidedly in favor of Catholicism, was replaced by Mary and William of Orange, the Dutch king, by an act of Parliament. William accepted the throne and furthermore accepted new laws passed by Parliament that barred the king from suspending any laws, raising any taxes, or maintaining an army without consent by Parliament. The new laws also protected citizens from arrest and detention without due legal process.

The ascendancy of Parliament in English political history ushered in an era of sustained economic growth. The Parliament was dominated by land-holding aristocrats who favored the expansion of property rights that benefited them. Though initially these rights were not in accord with the interests of the urban merchant class, that soon changed when land holders and merchants banded together in 1694 to lend the government money to pursue its war against France. Thus was created both a broad-based interest in cooperating across class lines and a precedent of government borrowing that made the English government by far the most consistently resourceful in Europe.

The English and the Dutch experiences, however different they were from each other during the sixteenth and seventeenth centuries, dovetailed in crucial ways. Both nations invested the power to tax in a representative body, and though neither Parliament nor the Estates-General was democratic in the contemporary sense, they effectively limited the power of the monarchy. The classes that gained the greatest benefits of legislative dominance — merchants and landowners — were the very classes whose interests lay in expanding private property rights and limiting taxation and government control over the economy. Consequently, one of the long-term effects of legislative dominance in England and the Low Countries was the gradual emancipation of economic energies from medieval strictures without the imposition of similarly inefficient royal strictures as in France and Spain. Eventually, the British and Dutch systems became models for development in other parts of Europe. To the extent that the core economic and political institutions that stimulated economic growth in Britain and the Low Countries were replicated, prosperity followed. However, when the seventeenth century drew to a close only a small fraction of the economic gains that private property rights, limited taxation, and legislative supremacy made possible had been realized. It was in the eighteenth and nineteenth centuries that the full potential of these institutions became evident in what we today think of as the Industrial Revolution. Before discussing the vast changes that occurred during the Industrial Revolution, however, it will be useful to briefly examine the economic and social institutions that emerged in Europe prior to the growth of industrialism in the eighteenth century. The institutional changes that made commercial activity central to European society in the period leading up to the Industrial Revolution were indeed as important as the revolution itself, for without them no great innovations in technology and production would have been possible.

6.3   Preindustrial commercial institutions

6.3.1   Legal Foundations of Capitalism

The legal foundations for a modern, commodity-intensive capitalist economy were laid long before the technological and organizational means existed to take full advantage of them. It might reasonably be said that the first legal reform that made possible the evolution of a capitalist economy was made in the Magna Carta in 1215, which decreed:

> No freeman shall be taken, or imprisoned, or be disseized of his freehold, or liberties, or free customs, or be outlawed or exiled, or any otherwise destroyed, nor will we pass upon him or condemn him, but by lawful judgment of his peers or by the law of the land.[3]

The Magna Carta was the first attempt to legally restrain the power of kings, and though the liberties that it protected pertained only to persons whose liberties were granted by the king in the first place, the first essential step had been taken in securing the rights of all persons to hold and trade property as they saw fit. Thus, from the late Middle Ages on, the concept of liberty was bound with some conception of economic freedom, freedom to own, buy, and sell property.

The second step was the creation of a forum for the adjudication of disputes between "freemen" (read the aristocracy) and the king, or between freemen and each other. These earliest courts dealt with economic issues among other things but were not commercial courts per se in that the complexities of contract law would still be centuries in the making. But these precommercial courts that negotiated the rights of royalty and the rights of freemen within a growing legal framework were a necessary beginning leading to the far more specialized and rationalized courts found in Antwerp, London, and elsewhere in the sixteenth century and thereafter.

The early courts in England were staffed by justices appointed by the king and were from the start concerned to prevent interference in what we now think of as free trade. Commons (1968) cites a case in which a court in 1300 fined three candlemakers for fixing their prices for candles among each other. These earliest courts established the principle of common law in England, and had a profound effect on the development of the concepts of liberty and property. Thus by the time John Locke considered the role of the free individual in society several centuries later, he could confidently declare that:

> The great and chief end ... of men uniting into commonwealths, and putting themselves under governments, is the preservation of their property.... The power of the society or legislative constituted by them can never be supposed to extend farther than the common good, but is obliged to secure every one's property.... And so, whoever has the legislative or supreme power of any commonwealth, is bound to govern by established standing laws, promulgated and known to the people, and not by exemporary decrees, by indifferent and upright judges, who are to decide controversies by those laws.... And all this to be directed to no other end but the peace, safety, and public good of the people.[4]

Locke thus links the concept of liberty with the protection of private property rights, and makes this the highest "good of the people" that government is organized to ensure. The common good, then, in this conception is the protection of individual liberties, that is, the rights to life and property. Community for Locke is possible because individuals make a contract with each other for mutual protection against each other.

The evolution of legal thinking and institutions since the Magna Carta in England has led quietly toward an interrelationship between what we think of as liberty, justice, community, and rights in property. The adjudication of commercial disputes then must be understood not as peripheral to issues of justice in modern capitalist society but as perhaps the most central issues through which the rights and purposes of persons are negotiated within the legal context. The emergence of strictly commercial courts that adjudicate contracts and other business related disputes are therefore an outgrowth of the common courts.

Rosenberg and Birdzell (1986) point out another feature of Western law first clearly articulated by Weber in his comparative sociological studies. Weber argued that one of the laws inherited by the Western world from the Romans is based on strictly rationalistic principles and a logical form of argumentation. This contrasts with the many cultural, religious, "magical" considerations other forms of justice have incorporated into legal reasoning. The absence of such considerations in Western law makes possible the predictable, calculable, and legally enforceable rights in property that underlie the capitalist system.

In China it may happen that a man who has sold a house to another may later come to him and ask to be taken in because in the meantime he has been impoverished. If the purchaser refuses to heed the ancient Chinese command to help a brother, the spirits will be disturbed, hence the impoverished seller comes into the house as a renter who pays no rent. Capitalism cannot operate on the basis of a law so constituted. What it requires is law that can be counted upon, like a machine; ritualistic–religious and magical considerations must be excluded.[5]

Weber argued that the legal framework of capitalism comes prior to the technical and organizational innovations that we most closely associate with the kind of entrepreneurial and corporate capitalism practiced today.

> [M]odern rational capitalism has need ... of a calculable legal system and of administration in terms of formal rules. Without it adventurous and speculative trading capitalism and all sort of politically determined capitalisms are possible, but no rational enterprise under individual initiative, with fixed capital and certainty of calculations. Such a legal system and such administration have been available for economic activity in a comparative state of legal and formalistic perfection only in the Occident.[6]

Weber, like Locke, links the rights of the individual to initiate economic action and own the rewards thereof with a legally defensible conception of liberty, though unlike Locke, Weber provides only hesitant normative support for his observations. However, Weber provided the first empirical support for the notion that the legal system peculiar to the West was invested with the potential to allow the expression of the kind of human activity and energy that would lead to sustained economic growth and innovation. The development of the particular articulation of contract and other commercial law follows rather naturally from the initial principle that individual persons have the near absolute right to private property legally obtained, and that it is the purpose of the state to guarantee this right of individuals through all of their dealings with each other.

6.3.2   Bills of exchange, insurance, and double-entry bookkeeping[7]

The actual physical possession of money was necessary to complete a transaction throughout the ancient world and in the Medieval world up until the thirteenth century. The limits of this system are apparent, especially when one considers that the invention of paper money was still centuries away. The simple necessity of having to carry around bags of gold, silver, or other coins was enough to create a strong incentive for brigandage and to discourage large-scale transactions. The invention of the bill of exchange by Italian merchants circumvented this problem and other ones besides, notably the problem of collecting interest amid the still widespread prohibition against "usury" imposed by the Church. A bill of exchange is basically nothing other than a check, a promise that the bill can be used to draw on an account. Bills of exchange were first used with accounts that merchants held with each other, but as banking developed, funds could be drawn from central locations. The invention of the bill of exchange and deposit banking are closely linked, as Rosenberg and Birdzell explain:

> As bills of exchange came into wide use, lesser-known merchants began to deposit funds with more widely known merchants, in order to place themselves in a position to pay by bills of exchange drawn on the more widely known merchants. It did not take long for the merchants who accumulated these deposits to discover that only a small portion of the deposits needed to be kept on hand to cover withdrawals, and that the balance could safely be used to buy bills of exchange at a discount — that is, for lending money at interest despite the prohibition of usury. They thus introduced deposit banking as a profitable and growing business in a society that prohibited the payment of interest.[8]

In the bustling trading markets of Antwerp and Amsterdam, separate markets that traded bills of exchange were developed in order to supply other markets with an easily available source of credit. The availability of credit, in turn, further fueled the growth of trade in the Low Countries.

The invention of insurance further extended the productive possibilities of economies by minimizing loss through risk distribution. The first insurance was used to insure maritime trading voyages. Insurance "made it possible for merchants to venture increasingly large amounts of capital on the commercial outcome of a voyage without subjecting themselves to the less calculable uncertainties of the sea."[9] At the same time, insurers could distribute the risk of having to pay the policy among themselves; the greater number of insurers, the less the risk to any single one of them. With enough policies in hand, however, individual insurers were likely to reap significant profits even if some ships were lost to storms or pirates. The growth of the marine insurance industry was the first lesson that Western economies learned about minimizing risk, and thereby increasing entrepreneurial action, without proportionately raising costs.

Double-entry bookkeeping represented the formalization of the break between the association of the individual or family and the business enterprises run by them. Double-entry bookkeeping, the recording of gains and losses in separate columns, was initially invented simply to keep more accurate records — if the two columns do not add up, one knows that somewhere one has made a mistake. However, the system also forces one to think of the enterprise as an entity in itself, "either as a debtor to its owners or as itself the owner of its own net worth."[10] In the accounting ledger, the enterprise appears as an abstraction separable from all persons associated with it. This move was a necessary one that made later accounting and legal institutions that protected owners and enterprises from each other possible.

6.3.3   The "Spirit of Capitalism"

The commoditized growth economy is not possible without a cultural foundation on which to thrive. The medieval Christian culture contained many beliefs and practices that made the development of capitalist enterprises impractical and difficult if not impossible. The prohibition against usury is merely a specific example of a worldview that placed a relatively low value on economic productivity as against other values. Medieval Christianity, though by no means unitary and uncontested, was a part of the larger feudal system of governance. It reinforced that system by preaching the importance of stability within a well-defined social hierarchy. Social innovation was often opposed by the Church, because of potential threats innovation might pose to the social order in which the Church retained final authority over important spheres of life, especially morality but also including political allegiance. However, as many analysts since Weber have argued, the economically stifling beliefs and practices of the Church are not inherent in Christianity but in the particular kind of Catholicism practiced during the Middle Ages — a belief system that valued piety over success, stability over innovation, community over individual. This belief system came under direct assault during the Protestant Reformation, which set the stage for the development of a progressive, more individualistic ethic that was inherently more friendly toward the development of capitalist institutions.

In his famous *The Protestant Ethic and the Spirit of Capitalism,* Weber argues that certain specifically Protestant beliefs underlay the individual motivation of Protestants to engage in worldly activities that led to the creation of a fully capitalistic economy, even if that was not their stated intention. This is not to say that Weber believed that Protestantism alone was responsible for the rise of capitalism, but rather that Protestantism facilitated and sped up the development of capitalism by providing individual Protestants with the incentive to act in ways that their Catholic forebears, countrymen, and fellow Europeans had little inclination to act. Indeed, the countries that first achieved economic success, the Low Countries and Great Britain, were Protestant countries, while France and Spain, great powers both, were Catholic. Even today, the most Protestant parts of Europe are the most developed, and the most Protestant parts of individual countries like Germany are more developed economically than the Catholic parts.

By "spirit" of capitalism Weber is deliberately avoiding making direct reference to capitalist institutions, the causes of which he locates elsewhere in society.[11] Rather, Weber wants to point out that a specific moral attitude lies behind the actions of individuals, and that in the aggregate certain moral attitudes are more likely to lead to certain kinds of social phenomena. The spirit that Weber identifies as particularly capitalistic relies on an abstract valuation of material success for its own sake — the making of money and more money without regard for rational consideration of sufficiency.

> [T]he *summum bonum* of this ethic, the earning of more and more money, combined with the strict avoidance of all spontaneous enjoyment of life, is above all completely devoid of any eudaemonistic, not to say hedonistic, admixture. It is thought of so purely as an end in itself, that from the point of view of the happiness of, or utility to, the single individual, it appears entirely transcendental and absolutely irrational. Man is dominated by the making of money, by acquisition as the ultimate purpose of his life. Economic acquisition is no longer subordinated to man as a means for the satisfaction of his material needs. This reversal of what we should call the natural relationship, so irrational from a naive point of view, is evidently as definitely a leading principle of capitalism as it is foreign to all peoples not under capitalistic influence. At the same time it expresses a type of feeling which is closely connected with certain religious ideas.[12]

Weber argues forcefully that the acquisitive ethic of capitalists is not inherent in the nature of human being, but rather must be learned and integrated into the personality. "A man does not 'by nature' wish to earn more and more money, but simply to live as he is accustomed to live and to earn as much as is necessary for that purpose."[13] Weber identifies this ethic as "traditionalism" of the sort associated with the medieval Church and society. In order to instill men with the spirit of acquisitiveness, then, there must be some moral motivation, something to still their souls in order to bring forth the great effort of will and imagination necessary to make them work not for the sake of happiness or comfort but simply to satisfy an abstract desire to make money. The motivation that Protestantism provides is the notion of the calling to make money linked with the notion of eternal salvation.

The idea of a calling is a fundamentally religious idea. It refers to a divinely inspired task on Earth. Weber found that neither the classical civilizations nor the Catholic ones had any clear notion of a calling for worldly activity, as opposed to direct service to God as a member of the clergy. The notion itself is an invention of the Reformation and is defined as "the valuation of the fulfillment of duty in worldly affairs as the highest form which the moral activity of the individual could assume."[14] Luther himself articulated the idea of the worldly calling, but it was only under non-Lutheran Protestantism that Luther's calling became a functional catalyst of capitalism.

The impetus to perceive a calling for worldly affairs was greatly enhanced with the development of the ascetic branches of Protestantism such as Calvinism and Methodism. Calvinism especially had far reaching effects on both European and American capitalism. The Calvinist doctrine that only a small proportion of people are destined (preordained) to be admitted into heaven led Calvinists to search for earthly signs that they were among the chosen few. As divine grace was not something that could be earned but rather had been determined in advance, individuals might be quite naturally anxious to ascertain their fates. Worldly success was taken as a sign that God's grace was upon one, which, ironically, provided individuals with maximum motivation to work hard in order to see if their efforts paid off in the kind of success that would reveal evidence of God's grace. This kind of effort was entirely divorced from any motivation to seek pleasure; rather, the reverse was true. Worldly effort was understood to be a kind of service to God, a complementary doctrine to the doctrine of grace. By following a worldly calling Calvinists were not only looking for signs of their own salvation but were also making good on their obligation to serve God by following His directive to them personally. Thus, Protestantism, and Calvinism in particular, had the unintended consequence of fostering the kinds of activity that were most conducive to the early formation of a capitalist economy, namely, those that aided in the accumulation of capital. At the same time that Calvinists had every motivation to work hard and succeed in worldly affairs, they had no incentive to consume the fruits of their labors, as self-indulgence was considered sinful and would also diminish those very assets that were thought to be evidence of eventual salvation.

By the time that capitalist economies matured, the religious components of the early Calvinist business pioneers became irrelevant to the continuation of capitalist institutions once they were firmly established, and the ascetic component especially has become even undesirable in a consumer-demand-driven economy. However, even stripped of its religious origins, the spirit of capitalism remains a vital moral/cultural part of the capitalist system.

> The capitalistic system so needs this devotion to the calling of making money, it is an attitude toward material goods which is so well suited to that system, so intimately bound up with the conditions of survival in the economic struggle for existence, that there can to-day no longer be any question of a necessary connection of that acquisitive manner of life with any single Weltanschauung. In fact, it no longer needs the support of any religious forces, and feels the attempts of religion to influence economic life, in so far as they can still be felt at all, to be as much an unjustified interference as its regulation by the State. In such circumstances men's commercial and social interests do tend to determine their opinions and attitudes. Whoever does

> not adapt his manner of life to the conditions of capi-
> talistic success must go under, or at least cannot rise.
> But these are phenomena of a time in which modern
> capitalism has become dominant and has become
> emancipated from its old supports.[15]

Protestantism, especially in its ascetic manifestations, created the cultural context in which the accumulation of capital could be justified, indeed extolled. The unintended effects of this development have been far reaching, and have not been confined to Protestant countries for many years now. However, as we pointed out above, the legacy of the Protestant ethic can be seen in the still greater wealth of the dominantly Protestant countries, though that ethic is no longer necessary to motivate people to pursue wealth for its own sake.

6.4   The rise of industrialism and the modern economy

6.4.1   The context for the Industrial Revolution

The "Industrial Revolution" refers to the massive replacement of human and animal energy with mechanical energy that took place in Great Britain during the late eighteenth and early to mid-nineteenth centuries. Prior to this time, the great majority of work was done by hand, that is, using the energy of human and animal bodies with the help of hand tools of relatively primitive design. We usually credit the Industrial Revolution with initiating the changes that have led to our modern, machine-dependent way of life. However, at an institutional level, the changes that took place during the Industrial Revolution were only a continuation of a process that had begun much earlier, especially the revolution in agriculture that happened as a result of Parliament's victory over the monarchy in their battle for political supremacy.

After the Glorious Revolution and the entrenchment of landed and merchant interests in the now-ascendant Parliament, legislators sought to improve their own lots and those of their class brethren by opening the rights of property. Nowhere was the change toward a free market system of property rights more dramatic than on the agricultural lands. Traditionally under medieval custom enshrined in common law villagers shared rights in land with the large landowners whose fields they cultivated. The village system was based on open fields, common lands, and a quasi-collective form of cultivation. It was a labor-intensive method and formed the backbone of the rural economy and culture among the common people. It was, moreover, a stable system in that very few innovations had occurred during the course of centuries, lending the British countryside and culture an air of permanence. In current parlance, we might say that the system, however class-based it might have been, was sustainable. It was also very inefficient in purely economic terms.

Among the changes that Parliament made to property rights in land after the Glorious Revolution was the passage of acts of enclosure, which allowed landlords to fence formerly open fields. Consequently, common property resources were gradually transformed into privately held resources, disposable as landlords saw fit. Where once entire villages shared pasturage, paying fees in kind to titled owners, landlords now experimented with new forms of stock raising. Where once villagers planted their crops, landlords now experimented with new varieties and new, capital-intensive forms of cultivation. Indeed, the agricultural revolution taking place in Britain at this time was a combined technological and institutional revolution that in a short period of time radically transformed the British countryside.

Of the new techniques introduced by the wealthy landowning class was an expanded use of animal fertilizer, crop rotation, new farming implements, new crop varieties, and new breeds of cattle and sheep. As a result, British food production rose while the agricultural labor force, as a percentage of the population, sank. From an economic point of view, this meant that a great many people were released from agricultural labor and became available for other kinds of work. At a human level it meant that these people were displaced, finding themselves suddenly without work, food, or any guarantee that they could find either to sustain themselves. In other words, the agricultural revolution produced the wage-earning worker class — defined by material insecurity and mobility, the new English worker would, when the time came, make up a commoditzed labor force necessary to man the Industrial Revolution. In the meantime, they took work where they found it, on the farms as day laborers or as workers in the small-scale enterprises that flourished in the towns and cities.

In addition to the revolution in agriculture, the growth of markets forms an essential part of the context for the Industrial Revolution. For the British, the huge increase in the size of markets for goods and services came with the establishment of colonies in the Americas, the building of the world's largest merchant marine, and naval supremacy. British producers in the eighteenth century were faced with terrific new opportunities to sell their wares, but first they had to produce more of them. The growth of the markets in goods not only put pressure on the government to eliminate restrictive laws that limited the mobility of the poor or prevented enclosure of fields, but also put pressure on producers to move away from the system of craft labor with its vertical integration and toward mass production and specialization.

North (1981) argues that the technological innovations that define the Industrial Revolution followed rather than caused the factory production system that we associate with industrialization. In other words, the increased size of the markets drove merchant-manufacturers to centralize their production due to the efficiencies inherent in having specialized workers all working on a single product under the same roof. These efficiencies are mainly a function of the close supervision of workers that this system makes possible. Merchant-manufacturers therefore lowered their transaction costs by improving quality control of the goods that they produced. In short, by centralizing workers, manufacturers were able to impose uniformity of quality.

This form of work organization produced a secondary benefit, however, which in hindsight seems to be a natural part of industrial manufacturing — the invention of machines to assist laborers in their work or replace them altogether. The increased monitoring of workers made it possible to calculate the individual contributions that types of workers made to the production process, thus reducing the cost of imagining and inventing machines to replace manual labor. In other words, as the production process was broken down into specialized components, and all component organized into a fully integrated assembly, overseers could make a more rational calculation of the efficiency of each stage of production and tinker with ways of increasing that efficiency. The heart of North's argument, then, is that the innovation in work organization was a necessary predecessor to technological innovation. That said, once some of the main technologies that characterized the Industrial Revolution were invented they brought about social and organizational changes in their own right. In the end, the highly commoditized industrial economy that emerged during the nineteenth century in Great Britain depended on a congeries of factors, including organizational and technological innovation as well as the gradual refinement of property rights institutions and other rules and mechanisms that created markets and made them more efficient.

6.4.2   The Industrial Revolution

The Industrial Revolution combined knowledge and technics with capital and property rights in a way that had never before been possible or imaginable. Centered primarily in Great Britain but following closely in Germany, Northern Europe, and North America, the Industrial Revolution was fundamentally a revolution in commoditization. Indeed, the mobilization of existing and latent energies for the purpose of expanding the commoditized economy became a central goal of the European and North American nations and in time became a defining characteristic of modernity itself. Other aspects of what we think of as modern society — mobility, individualism, increasing wealth, secular instrumentalism — are all related in direct or indirect ways to the evolution of the commoditized economy.

In order to take advantage of the enlarged and emerging global markets for goods and of the production possibilities that the centralized workplace had started to offer, manufacturers needed more capital than they had previously operated on, which meant that they needed access to credit. By the nineteenth century, banking was a significant industry in Great Britain, and the institution of the stock company had been invented. Currency itself had become commoditized through the invention of the gold standard, making it possible to exchange currencies and move large amounts of money around the world. The revolution in agriculture made large landowners wealthier than they had ever been, and they invested a part of their profits in urban manufacturing companies. Through their Parliamentary bond and increasingly through investment and a common economic destiny, the landowning and the merchant class in Britain merged their interests and coordinated their political and economic action. The industrial era thus began in a place that had already achieved a high level of wealth, institutional development and integration, and military might. Once the innovations in technics began, they never stopped.

The great technological innovations of the Industrial Revolution encompassed three basic forms of commoditization. The first was the creation of machines that displaced human labor. Some of the most important of these were the fly shuttle, the spinning jenny, the water frame, the cotton gin, and the assembly line. These inventions made the large-scale manufacture of textiles and other goods possible and were basically an intensification of a process of invention that had been going on for some time. The second form of commoditization has had an even greater effect on the transformation of society and nature — the discovery and creation of new forms of intensive, mass-producible energy sources. The most important early invention was the steam engine, which in turn made it possible to dig deeper for coal, iron ore, and other metals and minerals. The new energy sources also made new forms of transportation possible, the most important of which was the railway. The Watt steam engine was put to use in the first locomotives, providing the first-ever means by which large numbers of human beings could safely cross continents in astonishingly short periods of time. The final form of commoditization that defines the Industrial Revolution is what North calls the "transformation of matter." The newly emerging sciences of chemistry and physics made it possible to transform materials found in nature into other, manmade materials at a scale and diversity never before imaginable. Not least of these breakthroughs came in what we now know as the petrochemical industry, in which chemists and engineers not only devised new and efficient means to pump crude oil out of the ground, but also discovered processes through which that crude oil could be transformed into petroleum, still the world's most potent, cost-effective form of energy. In short order, petroleum itself was transformed into yet other materials such as plastics and pharmaceuticals.

The development of industrial chemistry and other applied sciences was a part of the larger phenomenon of the growth of the sciences during the Industrial Revolution. It was during this time that the sciences were devoted to the purposes of commoditization and went from being relatively esoteric forms of inquiry that had a greater impact on worldview than on material condition to being at the forefront of economic and social change. The mobilization of science for technological innovation and economic growth has proved to be perhaps the most important of all the changes that occurred in the West during the Industrial Revolution. The development of the scientific disciplines has been variously linked to a demand for certainty during periods of social instability,[16] the decline of Church authority, and the capitalization of industry and the resulting demand for technological innovations during the

Industrial Revolution.[17] Whatever the causes of the rise of science, the relevant fact for this study is that during the Industrial Revolution the sciences, including the newly emerging social sciences, were put to use to develop new products and processes and (in the case of psychology and sociology) to make factory labor more efficient and the urban working class less restive. The integration of the knowledge production process with technological innovation, commoditization, and economic growth has only become more highly developed over the course of the last century, until it is no longer possible to think of science as an autonomous intellectual enterprise.

The final set of innovations that occurred during the Industrial Revolution were institutional. These in turn are divisible into legal and organizational innovations. One of the organizational innovations has already been discussed: the centralization of the workplace. Other new forms of organization included, most importantly, the development of management-based organizations and bureaucracy. More and more during the Industrial Revolution, efficiencies in production and distribution were gained through the development of managerial specialties. Managers oversaw workers, finding new ways to improve worker discipline and efficiency. Managers also formed the linkages between organizations, rationalized and integrated productive activities, and invested organizational resources. In the new industrial context managers were increasingly important elements of the productive chain. In order to keep the expensive machines running at maximum capacity, companies found it increasingly expedient to run their own distribution networks, which were staffed mainly by managers.[18] The bureaucratic model of organization, with its specialization of functions and integration through hierarchy, allowed organizations to grow in size as well as diversity of activities.

The legal institutions that fueled the Industrial Revolution included some that more clearly specified property rights, and others that changed the way resources could be combined and allocated. Examples of the former include the refinement of patent law to protect as property the outcomes of innovation. Examples of the latter include laws that made corporations possible and gave them the legal status of persons.

In all, the technological, intellectual, and institutional changes that occurred during the Industrial Revolution brought together productive resources in combinations that had never before been attempted. The innovations in work organization and legal institutions made the development of new technologies possible in the context of growing markets that could absorb the mass-produced goods. The systematic utilization of knowledge for the purpose of inventing new products and improving productive processes has resulted in the greatest advances in productivity and technology in human history.

6.4.3   *The rise of the corporation*[19]

Corporations represent a form of collective action organized to achieve economic and possibly other ends. Corporations make possible actions and accomplishments that individual or small groups such as families could never, under any circumstances, achieve on their own. Indeed, corporate organization is at the heart of the productive achievements of the modern economy.

Corporations have their origins in the joint stock companies in Britain in the seventeenth century. These companies did not have royal charters, but rather issued stock certificates that could be freely traded to investors who were not authorized to make actual business decisions for the company. However, joint stock owners were responsible for the debts of the company, and the legal system had no way of recognizing joint stock companies as distinguishable from other actors, leaving the assignation of property rights ambiguous when it came to disputes. Joint stock companies also suffered from the legal tradition of not allowing corporate associations to persist unless they were chartered by the state. The joint stock company was thus limited in its ability to survive hard times and to maximize its productive potential.

These drawbacks to the joint stock company were eventually resolved through a gradual process of legal reform. By the mid-nineteenth century the corporation in Britain and the U.S. was recognized as having the legal rights of a person, which in effect insulated the officers and owners of the corporation from liability for its debts. This in turn freed corporations from constraints on risk-taking that were inherent in the system when the actual wealth of actual persons was at stake if the enterprise failed. In other words, corporations allowed a new and very powerful form of entrepreneurship to emerge. The modern corporation is certainly not the only kind of business enterprise to succeed in the capitalistic economy. However, it is clearly one of the most important, and shares with other forms of business organization the characteristic of joining the productive and creative capacities of many persons in a collective enterprise.

6.5 The American experience

6.5.1 Constitutional foundations and the legal impetus for development

The development of the highly commoditized economy in the U.S. is a place-specific continuation of the economic development that had been taking place in Europe in the early modern period. The first settlers from England happened to be precisely the kinds of people who were most likely to make the most of their inherent economic potential, Calvinist Protestants. Moreover, they brought with them the English system of common law that protected private property rights and limited the rights of government to interfere in the business affairs of individuals. The work ethic and thrift of the Puritans left later generations of Americans with a solid economic foundation upon which to introduce such innovations as the corporation and deposit banking and to apply knowledge and new technologies to productive enterprises. The Puritans left behind no less than a robust and consistently growing economy as well as certain cultural traits that, in diluted form, still permeate the character of American business.

By the time of the American Revolution, the country had long since enjoyed a fairly high level of development, and Americans understood themselves to be in the forefront of not only a political movement toward representative government but of an economic movement toward unprecedented wealth and the freedom of individuals to pursue it through entrepreneurial effort. The framers of the Constitution, for all of their differences of opinion and conflicts of interest, generally agreed with the principle articulated by Locke that the liberty of individuals consisted chiefly in their right to pursue their fortunes with as little interference from government as possible. Significant parts of the Constitution can be understood as attempts to institutionalize this principle. James Madison expressed this idea clearly in his now famous *Federalist 10*:

> The most common and durable source of factions has been the various and unequal distribution of property.... The regulations of these various and interfering interests forms the principal task of modern legislation, and involves the spirit of party and faction in the necessary and ordinary operations of government.... The interference to which we are brought is that the causes of faction cannot be removed and that relief is only to be sought in the means of controlling its effect....

Madison understood that economic factions will attempt to use government to redistribute wealth to benefit themselves, and therefore considered it the foremost duty of the constitutional framers to devise a political system that minimized such unjustifiable interference in the affairs of free men. The American Constitution was indeed an innovation in the organization of government, and embodied the principles of the Madisonian vision for a political system that allowed the creative and industrious capacities of its citizenry to develop unregulated by the state. The system of checks and balances — three branches of government with the legislature broken down into two chambers of differing institutional constraints and cultures — was meant to insure that no faction (a group of persons seeking to make policy to benefit their own specific interests) could easily prevail in dominating government. There are many more opportunities to fail in changing the American legal system than there are in making positive law. However, the Madisonian system was not suited to fit the social and economic developments that came during the eighteenth century, and the restrained government that Madison envisioned was eventually replaced by a more active government that sought to take positive steps to encourage economic development rather than just allow it to happen at the instigation of private parties only.

In the nineteenth century the courts were the first branch of government to develop common law that actively promoted economic growth. This was followed by statutory developments that set the rules for incorporation, insurance, credit, and banking. Nineteenth-century legal philosophy was oriented toward the "release of energy," that is, toward making it possible for individuals and groups to develop economic enterprises in an unfettered market.[20] The common law and then statutory thus established the framework for property rights that eventually became codified as the fundamental principles of the modern capitalist economy. The innovations in American law were not in their basis different from the innovations that took place in Europe. However, there were some specifically American developments that made forms of economic growth possible here that were not possible elsewhere. In particular, the great wealth of land and other natural resources found on the North American continent provided Americans with an opportunity for wealth creation that few other peoples enjoyed. The homesteading acts that distributed this land to the common people represents a uniquely American attempt to marry economic growth with popular democracy. The success of this effort is plain not only from the great productivity of private lands in the U.S., but in comparison to the land distribution and historical productivity of lands in South America. Though no less exploited for their economic potential, land in South America traditionally was distributed mainly among wealthy landowners, a "faction" in the Madisonian idiom. Not only do the South American systems interfere with the development of democratic institutions, they have severely inhibited the growth of national economies, though they have certainly made a small number of people very rich.

6.5.2 Reform and adaptation

Whatever its successes, however, as early as the late nineteenth century the costs of rapidly growing industrial capitalism had shattered the faith of many Americans in the absolute virtue of economic development. As opposed to the relatively democratic distribution of land in the U.S., the distribution of wealth in urban areas reflected the distribution of ownership over industrial resources. Which is to say that classical industrial capitalism was very skewed to favor the enrichment of a small number of persons, while the great many toiled for miserable wages. The social costs of industrial development were enormous and could be counted in foreshortened lives, child labor, and general urban poverty. The environmental costs were becoming a burden as well, as great swaths of country were denuded of forests for the benefit of timber barons and wildlife resources were decimated for the benefit of hatmakers who used bird feathers in their products and railroads that wanted to eliminate buffalo (and Indians) from the plains.

Reform of the capitalist system began late in the nineteenth century with laws that included child labor regulation, a limited workweek, meat safety regulation, drug regulation, and "rational" management rather than unregulated exploitation of natural resources. The rise of the labor union movement and the conservation movement around the turn of the century eventually helped bring economic practices under government regulation. The shift from a completely free market system to one of regulated capitalism in the long-term made the capitalist system more sustainable by minimizing social conflict and preventing the rapid depletion of resources. The changes in law made during the Progressive Era, especially under President Teddy Roosevelt, are now understood to be the first steps in creating what we think of as the regulatory state. The New Deal continued this evolutionary trend to restrain the destructive effects of unregulated capitalistic development.[21] Likewise, the concern of this book, checking commoditization and its destructive effects on society and nature, are a continuation of the effort to regulate capitalism in order to make it serve society rather than the reverse.

Notes

1. This section is drawn primarily from North, D., *Structure and Change in Economic History*, W.W. Norton, New York, 1981.
2. North, D. and Thomas, R., *The Rise of the Western World: A New Economic History*, Cambridge University Press, Cambridge, 1973.
3. Adams, D., *Renewable Resource Policy*, Island Press, Washington, D.C., 1993, p. 21.
4. Locke cited in Morris, C., *The Great Legal Philosophers: Selected Readings in Jurisprudence*, University of Pennsylvania Press, Philadelphia, 1978, p. 152.
5. Rosenberg, N. and Birdzell, L.E., Jr., *How the West Grew Rich: The Economic Transformation of the Industrial World*, Basic Books, New York, 1986, p. 116.
6. Weber, M., *The Protestant Ethic and the Spirit of Capitalism*, Charles Scribner's Sons, New York, 1958, p. 25.
7. As the discussion in this chapter shows, particular institutions arise out of particular historical circumstances, and cannot be reduced to a single cause, especially not to a cultural phenomenon such as religion. The rise of the sciences and the nation–state, the power of legislatures and the taxation policy operative in particular countries accounts sufficiently for economic institutions such as the firm and factory.
8. Rosenberg and Birdzell, *How the West Grew Rich*, p. 117.
9. Rosenberg and Birdzell, p. 118.
10. Rosenberg and Birdzell, p. 127.
11. Weber, *The Protestant Ethic*.
12. Weber, p. 53.
13. Weber, p. 60.
14. Weber, p. 80.
15. Weber, p. 72.
16. Toulmin, S., *Cosmopolis: The Hidden Agenda of Modernity*, University of Chicago Press, Chicago, 1991.
17. Musson, A.E. (Ed.), *Science, Technology, and Economic Growth in the 18th Century*, Columbia University Press, New York, 1972.

18. Chandler, A., *The Visible Hand*, Belknap Press, Cambridge, MA, 1977.
19. This section is based on Rosenberg and Birdzell, *How the West Grew Rich*, Chapter 6.
20. Hurst, J.W., *Law and the Conditions of Freedom: in the 19th Century United States*, University of Wisconsin Press, Madison, 1956.
21. North, *Structure and Change.*

chapter seven

Ecology and commoditization

Contents

7.1 Sustainable development and the challenge of ecology

We certainly know enough now about ecology, and we are clever enough about the workings of the world to know the sorts of changes we need to make in order to create a more ecologically sound and sustainable way of life. We have known for some time that humans, like all life forms, extract resources from our environment and return wastes. If we take faster than the Earth can give and/or leave more than the earth can process, the balance of nature teeters and our lives become more difficult and less stable.

For development to be ecologically sustainable, knowledge gained from careful study of the impact of human activities on the health and functioning of ecosystems must be fed back into the development process and used to adjust those activities. The capacity to learn and adapt is the starting point for sustainable development. Sustainable development requires a change in the relationship between human beings and the biosphere, a change from a relationship of parasitism to one of symbiosis. In a parasitic relationship, one party prospers at the expense of the other; in a symbiotic relationship, each enhances the other's capacity to thrive. To consciously design our communities, industries, and ways of life so as to benefit the natural world should be the highest goal of sustainable development. Short of that, we can at least try to do less harm.

Ecology might be the most politically challenging of the sciences, because its discoveries suggest that contemporary modes of human living cannot survive long into the future. Ecology also gives us the basic tools with which to consciously redesign our ways of life to disrupt the earth's natural cycles far less than we do today. However, the lessons from ecology fly in the face of the institutions and practices that maintain an economic system driven by commoditization. While commoditization virtually demands that more and more of the Earth's resources be wrenched from their tangle of coevolved ecological relations, ecology shows how this may lead to long-term losses in ecosystem productivity and eventually in ecosystem collapse. We may never fully grasp how utterly dependent we are on healthy ecosystems until they start to unravel. This lesson has not been lost on people in the Sahel and other places where the basic structures and functions of ecosystems have been destroyed. The problem of commoditization must be addressed or all the well-intentioned efforts to make modern economies more sustainable and environmentally benign will inevitably fail.

7.2   *Ecological principles and economic implications*

As long as we rely on market-based decisions to determine how we allocate the vast majority of our time, attention, and resources, then the commoditization forces I have described will accelerate and will overwhelm efforts to build a sustainable world. In the end, our environmental fate will be determined by how successful we are in developing policies that provide counterbalancing pressures to commoditization. Decommoditization by definition leads to less economic growth, even economic contraction. Richard Douthwaite, Herman Daly, and others have convincingly argued that we need to move beyond growth to a steady state economy.[1]

The principles most relevant to an ecological understanding of economics are also those upon which a "sustainable" economy could be built. The remainder of this chapter will present each of these principles in general terms, the implications for environment and society, and how commoditization distorts economic behavior so as to place modern human societies out of sync with these principles:

1. Economic systems are subsets of ecological systems: The principle of ecosystem primacy.
   Implication: Ecological considerations should trump economic ones.

2. Energy is the primary natural resource: The principles of entropy and conservation.
   Implication: That is best which wastes energy least.
3. Efficiency is enhanced by working with natural flows and processes rather than against them: The principles of appropriate technology and ecosystem thinking.
   Implication: Technology should be designed to work with rather than against natural flows of energy and materials.
4. Contradictory goals cannot be maximized at the same time and must be balanced: The principles of homeostasis and optimality.
   Implication: Information indicating that the human economy is out of balance with nature must be received and processed and adjustments made to optimize the sometimes conflicting goals of prosperity and ecological integrity.
5. Scale and level of organization matter: The principle of cooperative hierarchical organization.
   Implication: Economic policy decisions should simultaneously consider effects at the level of the individual, the level of the economic system, and the level of the global ecological system.

These principles can be considered allocation principles, determining where we should allocate human attention to live more compatibly and lightly on the earth. The pressures of commoditization make it difficult if not impossible to live by these principles. Because of commoditization economic goals almost always trump ecological ones. It is the price of fuel, not the logic of entropy, that determines how readily energy will be wasted. Working with natural flows requires long, careful observation and site-based ingenuity, neither of which can be mass-marketed. Balancing goals means ending the primacy of economics, and in effect ending the power of commoditization to allocate human attention. And taking system-level considerations into account when making economic decisions means giving processes equal value as products, and this, as we have seen, is contradictory to the imperatives of commoditization. The following sections go into more detail about each of the principles and their implications as well as the barriers commoditization places in the way of achieving an economy based on these principles. This will lead to Chapter 8, which will propose policies for countering the excessive power of commoditization in order to build an economy consistent with ecological principles.

7.2.1   Economic systems are subsets of ecological systems: the principle of ecosystem primacy

Implication: Ecological considerations should trump economic ones.
People and societies are not exempt from the laws of nature. And yet we think and behave as if these laws have no ultimate meaning to us. The sheer abundance of our species is evidence that we have managed to postpone

reckoning with the rules of ecological carrying capacity. An ecosystem's capacity to support, or carry, a given population of animal or plant is limited by available resources and the complex dynamics of ecosystem balance. Human intelligence and creativity have made it possible to utilize a far wider range of resources, and to obtain resources from above and beneath the land and oceans. Through our tremendous advances in transportation we have learned how to draw materials and energy from hinterlands far distant from our concentrated settlements, our cities. Our ingenuity makes it possible to find new sources of raw materials, to make synthetic substitutes when nature grows stingy, to continually get more out of lower quality raw materials and energy as the higher quality sources are depleted.[2]

Humans not only place demands on the environment through our place at the top of the food chain, but also have created an immense economic organism with its own highly organized process of appropriating materials (or nutrients) and secreting wastes, in effect an economic metabolism. When we begin to think of the human economy as an organism whose metabolism places demands on its environment we can no longer avoid the implications of ecological limits. Probably one of the most compelling shifts of perspective was when Herman Daly placed the familiar box diagram and model of the economy inside a larger box representing the Natural World (Figure 7.1). According to Daly, the old model illustrated a belief "that the economy is an isolated system in which exchange value circulates between firms and households. Nothing enters from the environment, nothing exits to the environment.... For all practical purposes, an isolated system [that] has no environment."[3]

When you look at Daly's illustration, three things stand out as obvious. First, the Natural World is the source of all materials for the economy (Daly and other ecological economists like to call this Natural Capital). Second, all the waste products of the economy are returned to Nature (in degraded condition). Third, the economy can only grow so large before it begins to fill all the available natural space. William Rees summarizes it neatly: "The material economy is an integrated, completely contained and wholly dependent growing subsystem of a non-growing ecosphere."[4]

Rees of the University of British Columbia and Mathis Wackernagel at Universidad Anahuac de Xalapa in Mexico have created a tool, ecological footprint analyses, that takes the notion of carrying capacity and makes it meaningful to the human economy in a compelling way. As we noted previously, industrialized and urbanized societies have so far been able to postpone a reckoning with the implications of limits to carrying capacity by being able to import resources and export wastes from distances far removed from their immediate settlements. This has been facilitated by tremendous advances in packaging and transport associated with commoditization. Rees and Wackernagel's ecological footprint analysis measures for any given population the area of land and water required to service that population's economic metabolism, i.e., produce the resources consumed and assimilate the waste produced by that population. Different analysts have used a somewhat different

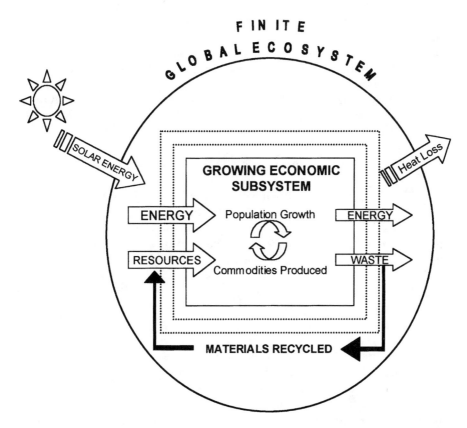

Figure 7.1 Commoditization accelerates the growth of the economic subsystem. It selects those goods and services with high embodied material and energy throughput (both energy and material input and waste energy and pollution output). (Adapted from Daly, H., *Steady State Economics*, W. H. Freeman, San Francisco, 1977.)

approach to adding up the size of the footprint depending on what variables are considered, but all show that a typical wealthy urban area in the industrialized North impacts and degrades a far greater area of forest, agricultural land, ocean, lake and wetland than the area of land they actually occupy, around 300 to over 1,000 times greater.[5] The International Institute for Environment and Development, for example, studied the ecological footprint of the city of London and demonstrated that the amount of the Earth's surface required to maintain the city's economic metabolism was equivalent to the entire area of all of England.[6]

Rees estimates that "with prevailing technologies and average consumption levels, the present world population exceeds global carrying capacity by up to one third."[7] Since the wealthiest 20% of the world's population presently consumes 80% of the world's resources, it is difficult to avoid the conclusion that the wealthiest of the Earth's people have already appropriated more than the available carrying capacity of the planet, leaving virtually nothing to the remaining vast majority of the Earth's population. Others have calculated that from 40 to 50% of all the biological productivity is presently transformed or degraded by human activity. This figure shocked the world of environmental science when it was first published in *BioScience* in 1986.[8] More recently, Peter Vitousek and colleagues have summarized global signals that the impacts of human activities have begun to significantly alter several of the Earth's key ecological indicators,[9] including the following:

- 20% increase in atmospheric CO_2 concentration related to human activities
- 50% of all the accessible fresh surface water is being used
- Humans are responsible for over 50% of all terrestrial nitrogen fixation
- 20% of all current plant species in Canada are invaders from elsewhere, with similar percentages likely in other parts of the world
- 20% of all bird species on Earth are now extinct, mostly as a consequence of human activities
- 60% of major marine fisheries are considered fully exploited, overexploited or depleted

Herman Daly described the difference between what he called empty world economics and full world economics. This distinction and its social and political implications lie at the core of the challenge that ecological economics presents to mainstream economists. In a full or nearly full world, as ours apparently is, the dangers of continuing on a path of unfettered economic growth are enormous. As we approach these environmental limits to economic growth the associated costs increase. These environmental costs are largely ignored by our conventional economic accounts such as GNP. Given this, Daly suggests we may be entering, or have already entered, a period of "antieconomic growth" in which the actual costs of growth outweigh the benefits.

The only way to reduce or stabilize economic growth while maintaining a high quality of life is to increase the amount of service provided per unit of economic output. But the means to accomplish this through real conservation and increased community self-reliance, as we have seen, rely heavily on goods in the economy of care and connection, which have intrinsically low commodity potential. Furthermore, as long as market forces determine economic behavior, only that which can be priced, bought and sold matters. Free goods and common goods, no matter how important, are considered extraneous to the economy. In the logic of commoditization, the natural world is merely a storehouse of raw materials for the production of commodities. What's left out is everything that resists commoditization which are, as we have seen, processes rather than products. These include ecosystem processes which produce and clean the air and water and build the soil, the planetary metabolism which maintains climate and ocean stability, the biogeochemical cycles of the critical minerals and nutrients of the planet, and the processes of evolutionary change and genetic diversity. No matter that nothing is producible without it, nowhere will the economy value it as long as commoditization underlies the logic and practice of valuation. The model of an economy of firms and households abstracted from their environment is an illusion that confuses much more than it informs.

The problem is not with the discipline of economics nor with economists per se. They are doing their job as it is defined for them. The questions economists ask are subject to the same selection pressure of commoditization as everything else. The questions that survive and that matter in a commoditized economy are those whose answers inform the needs of managing or functioning within that economy. The tools of the economist are put to the service of the commoditized economy — that's who pays the bill. The tools economists have invented answer the questions about the flow and exchange of commodities. Only by political means can we assert values other than market values and so make it meaningful and rewarding to ask different questions of economists, including ecological economists.

In the logic of the argument of this book, ecosystem services are inherently services with low commodity potential: they are relational, local, and complex — the exact opposite of goods with high commodity potential, which are independent, universal, and simple. The economics of forestry can be reduced to the culturing, harvesting, processing, allocating, distributing, and storing of forest products. The complex dynamics of forest ecosystems and their roles in producing clean air and water and providing habitat for diverse forest life will only be protected and valued when such value is recognized as specifically noneconomic in a commoditized economy and is dealt with politically as a question of the common good.

Ecosystem services must be recognized as public goods that are to be protected by institutions with the capacity to protect the commons. This requires that governance evolve so as to gain the capacity to regulate the economy for specific environmental ends. This will be the fundamental principle in the design of policies to counter the effects of commoditization discussed in more detail in Chapter 8.

7.2.2   Energy is the primary natural resource: the principles of entropy and conservation

Implication: That is best which wastes energy least.

There are many excellent treatments of the implications of the first and second laws of thermodynamics for ecological economics. Daly offers the following useful summary. The first law suggests that energy cannot be created or destroyed, but only changed in form. The second law, also called the entropy law, suggests that the ability of energy to do useful economic work only decreases. Taken together, we only have so much useful energy available to us, and its ability to do work is in constant decline. Therefore, it is the availability of useful (low-entropy) energy that marks the fundamental limit of economic production. We currently have two very different sources of low entropy energy available to us. A solar source with an unlimited stock, but constricted flow, and a terrestrial source (fossil fuels) with a limited stock, but unconstricted flow. According to Daly, after relying on the solar source for much of human history, we have recently become addicted to the terrestrial source, and the economic growth it provides. We have switched our dependence from the unlimited to the limited source of low-entropy energy. Inevitably, we will be forced to again live within the constraint imposed by the daily flow of solar energy bathing our planet. The sooner we begin to make this transition, the smoother it will go, but it will be difficult because it flies in the face of commoditization.

There are many other excellent treatments of the second law of thermo-dynamics and the implications of entropy for ecological economics.[10] Paul and Anne Ehrlich and John Holdren summarize the meaning and implications of the second law in this way:

- In any transformation of energy, some energy is degraded.
- No process is possible whose sole result is the conversion of a given quantity of heat (thermal energy) into an equal amount of useful work.
- No process is possible whose sole result is the flow of heat from a colder body to a hotter one.
- The availability of a given quantity of energy can only be used once; that is, the property of convertibility into useful work cannot be "recycled."
- In spontaneous processes, concentrations (of anything) tend to disperse, structure tends to disappear, order becomes disorder.[11]

Suffice it to say that the production of goods and services through the transformation of raw materials into useful products requires more energy than can be embodied in the goods and services, or reused or recycled from them. The economy of goods and services necessarily degrades the resources that it draws upon. The production of goods is always accompanied by the production of "bads." The energy that drives the economy goes in one direction only: usable resources become more scarce and waste more abundant. Most environmental problems are traceable to this fact.

Energy is only useful to us economically if it can be channeled to do work, in other words be applied to matter in such a way as to cause physical or chemical change (or to maintain structure in the face of entropic change). It takes work to hold things together. Structure, whether it is a human body or a chair, requires the application of work. Work must be powered by energy. Constrained by the second law of thermodynamics, energy can never be totally transformed into work; some must always be dissipated as the lowest quality energy, heat. There is, however, considerable gains that can be made in improving the efficiency in which energy is used to produce goods and services, especially in the U.S. Figure 7.2 shows the slight downward trend in the amount of energy used per U.S.

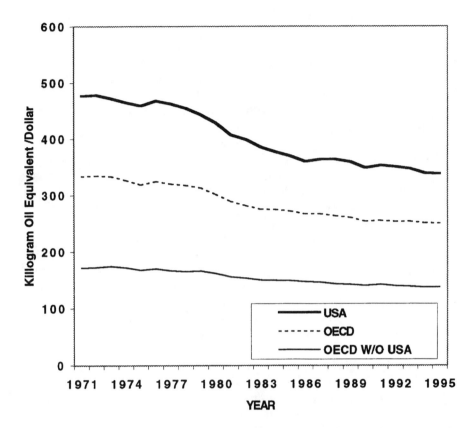

Figure 7.2   Total primary energy supply per one dollar domestic production: 1971 to 1995. (*Source:* United States Statistical Abstracts, 1973–1997 International Energy Agency, OECD Countries Energy Statistics, Annual.)

$1 of domestic production in industrialized countries. The U.S. uses more than three times the amount of energy for an equal amount of production than other developed countries.

Looking at energy we can see the difference commoditization makes. Assuming that energy efficiency makes sense, there are three possible approaches and technologies that could be developed, in descending order of commoditization potential:

- To develop and exploit ever higher quality fuels so that more of the energy embodied in fuel can be delivered at the point of production of a good or service and less is dissipated and lost (fuel efficiency)
- To reduce the amount of fuel required to produce any commodity (production efficiency)
- To organize human communities in such a way as to reduce or eliminate the need for the commodity (consumption efficiency or decommoditization).

Each approach requires the application of creative problem-solving and cooperative efforts, and each is capable of reducing the amount of energy used per unit of service, but each has not and could not receive equal amount of attention and research and development because of the differences in commoditization potential. The advances made in fuel and production efficiency have far exceeded the advances made in consumption efficiency.

That the greatest amount of effort has gone into improving fuel quality is consistent with the rule of commoditization. High-quality fuels are those that are most commercially useful, meaning they exist in concentrated forms, are relatively easy to store and transport, and can be most efficient and productive in powering work. In short, the higher the quality of a given fuel the more it contributes to overall commodity potential in an economy. Technological innovation has been directed toward making it possible to obtain and market fuels that pack a greater wallop per dollar invested. Thus the history of industrialization has been marked by transformations and improvements in fuel quality, starting with biomass fuels (mostly wood and peat and whale oil), then moving to coal once the technologies of mining and earth moving were developed, then to oil and gas once the technologies for drilling and refining were developed. Centralized electricity generation and distribution made it possible to deliver energy where and when it was needed at a considerable distance from the fuel combustion site.

Advances in production and transportation efficiency have led to reduced costs of production of commodities, making them increasingly less expensive and more developed than noncommodities. Since the invention of internal combustion engines, the history of industrial technology can be seen as the utilization of ever-increasing amounts of higher-quality fuels, at ever-increasing efficiencies to replace and/or supplement human driven work. Labor productivity increased more than a 100-fold since the industrial revolution by substituting energy for human labor. This substitution greatly enhances commoditization, because labor is far less commoditizable than energy. The labor released from production in a commoditized economy must seek out employment in commoditized service industries, which also expands commoditization there. Over time fuel replaces workers in these industries as well and another round of commoditization ensues. Without some counterbalancing decommoditization force that replaces energy with human labor and creativity the mobilization of energy and materials must continue with all the accompanying environmental and social costs.

Both fuel efficiency and production efficiency is consistent with the imperatives of commoditization and therefore have been subject to considerable development in modern economies. The third form of energy efficiency, consumption efficiency, is antithetical to commoditization and in fact might be called decommoditization. The tools and skills of energy efficiency at the point of consumption belong in the hands of the final user, not the energy producer. This path to efficiency has simply not been tried to any great extent.

The story of the centralized production and distribution of electricity is a good example of how development gets distorted by commoditization. The distribution of electricity through the power grid results in considerable losses in fuel-to-work efficiency (most plants operate at around 35% efficiency, meaning they produce two units of waste heat for every one unit of electricity produced, and even more is lost in distribution and end use). This efficiency is sacrificed for improvements in centralization of power and ownership, convenience and portability, qualities characteristically favored by commoditization.

While the commoditization of energy has driven enormous technological changes, there has been little in the way of comparable advance in the technologies of end-user efficiency. Such technologies include building design and location for passive solar heating and cooling, small-scale jerry-rigged windmills and farm-scale methanol production, small scale neighborhood energy storage, as well as many reduced consumption alternatives such as neighborhood equipment and tool libraries, energy-efficiency design cooperatives, and many others. Such advances require specific technical innovations at the point of use. Such technical advances are far less commoditizable because they necessarily involve decentralized, site-specific problem-solving. The point is not that these three energy paths — fuel efficiency, production efficiency, and end-use efficiency — are mutually exclusive, but that commoditization creates a severe imbalance in allocation of research and development resources so as to overdevelop certain aspects of energy production and use while underdeveloping key technologies that could dramatically reduce the amount of energy used to support a high quality of life.

This imbalance will only grow more important as we approach the limits to how much energy wastage the planet can tolerate. As readily available sources of high-quality fuel become more scarce, the amount of effort and energy required to obtain, process, and transport a given quantity of energy rises, thus threatening the continued advance in net energy productivity and increasing the amount of resource depletion and waste per unit of energy delivered. Energy in the forms of high-quality fuels capable of powering economic activity can only be used once. Every gallon of oil burned is permanently lost, while the supply of fossil fuel is 1 gallon diminished. Fossil fuel and other highly concentrated forms of energy resources are necessarily nonrenewable. As time goes on the highest quality and most easily extracted and processed fuels are used first. The cost in energy of producing more energy continues to rise. At some point the cost in energy expended is equal to or greater than the amount of usable energy obtained, at which point improvements in technology, or new sources or new fuels are needed to increase the energy return on investment or else economic advance must begin to slow and eventually stop. Other resources can also be evaluated for quality in terms of the amount of energy needed to transform the resource into an economically useful form. All of these analyses emphasize the absolutely critical role of energy availability and quality in all aspects of the human economy.

The evolution of human civilizations is sometimes described as the story of converting potential energy into useful forms. The story begins with the domestication of fire, then tells of the invention of the wheel and the harnessing of wind energy in ship's sails. With wheel and sail comes the power to move people, armies, and goods across land and water. These give military advantages in particular to those societies whose economy and technology become dependent on conquest and exploitation. As we saw earlier, commoditization took root in these societies and began to overdetermine technological evolution, beginning with the industrial revolution and the expansion of colonialism and imperialism.

The conundrum facing modern industrialized societies is this: increasing economic development and human prosperity has up until now been directly correlated with increasing energy use.[12] Most economic projections at the end of the twentieth and start of the twenty-first centuries assume that economic growth, particularly enhanced by the aspirations of the majority of the world's people to attain standards of living comparable to the industrialized minority, will lead to dramatic increases in energy consumption. With current trends these energy demands are likely to be met by nonrenewable fossil fuels, with all the accompanying increases in pollutants and CO_2 emissions.

We'll have to get a whole lot more efficient in our use of energy in order to avoid global environmental calamities. Reddy and Goldemberg (1990) have argued persuasively that it is possible to raise living standards in the South to a standard equivalent to that enjoyed in Western Europe in the 1970s with a flow of about 1 kW of energy per capita used continuously, a small fraction of the energy presently consumed per capita in the rich industrialized countries today. This level of energy efficiency, however, would require massive changes in patterns of energy use amounting to a decommoditization strategy that would rely heavily on production and end-use efficiency and on generating energy as synchronously as possible with the ultimate use. What Reddy and Goldemberg concluded will sound familiar to the reader by now. "A new paradigm for energy use is therefore essential. Energy must be viewed not as an end in itself or as a commodity but as a means of providing services."[13]

The laws of thermodynamics place real limits on improvements in energy efficiency. There are also serious and growing limits to the ready availability of high-quality fuels, which make efficiency improvements even more difficult to obtain. Although there is considerable room for improvements in energy efficiency, most of the energy-saving prospects, such as decentralization of energy production and technological advances in conductivity, require redesign and restructuring of current industrial and energy production and distribution facilities. The massive energy and material costs to carry out this restructuring is often not considered in the calculations of prospective energy savings. Lastly, without a strategy of decommoditization no amount of efficiency gains are likely to help. Two thirds of the energy used in industrialized countries like the U.S. is used in transportation and in home heating, lighting, and appliances. There are tremendous efficiency gains to be made in these areas, but they are inherently decentralized and labor-intensive and thus involve low commodity intense skills and services. As long as commoditization favors the centralized and the capital intense, these efficiency gains will be not live up to their potential. In an economy under heavy commoditization pressures what happens with a dollar saved when efficiency gains lowers energy costs is more than likely spent on another commodity, in effect increased efficiency leads to increased consumption.

This has been the case in the history of industrial development. Most of the efficiency improvements to date resulted from improvements in the quality of the primary fuel being used. This efficiency trend has been consistent with commoditization as high-quality fuels also have greater commodity potential. Each step in the technological progress of fuel from wood to coal to oil has seen an improvement in energy density and the capacity to store and transport energy. It's worth noting that each step also represents a decrease in the carbon/hydrogen ratio of fuel, meaning less CO_2 is emitted per unit of fuel burned. Yet despite these improvements, because of commoditized economic growth the total amount of carbon dioxide added to the atmosphere continued to rise dramatically throughout the evolution of efficient and cleaner fuels. The fact is, energy efficiency programs will only be successful in uncoupling improved quality of life from increased energy use if it is accompanied by a political and economic strategy to counteract the effects of commoditization. Without it, improved efficiency leads to lower costs and increased consumption. This paradox is sometimes referred to as "Jevons' paradox" after economist Stanley Jevons pointed out in 1864 that efforts to conserve English coal by increasing the coal-use efficiency of British steam production ended up making steam power cheaper compared to human and animal power and in the end stimulated increased coal consumption.[14] Fuel efficiency gains made in automobile engines have had similar effects. A study by Freund and Martin (1993) demonstrated that even though the efficiency with which automobiles used gasoline in the U.S. improved considerably (34%) between 1970 and 1990, total fuel consumption during the same period increased by 7%, because the number of multi-car families had increased and the family drove more miles.[15] Fuel efficiency gains have also been erased by the increasing size of the American car as sport utility vehicles and minivans have gained in popularity.

Many analysts have pointed out that in order to meet the need for economic growth in the Third World without dramatically increasing the amount of global environmental stress, there needs to be an order of magnitude (a factor of 10) improvement in the amount of material and energy used per unit of service provided.[16] Such efficiency improvements, while theoretically possible, cannot merely lead to producing more goods and services with less energy per unit, but also must satisfy human needs with fewer goods and services. This, as we have seen, is opposite of what happens as a result of commoditization.

How dependent are we on increasing energy use to achieve and maintain prosperity? The data are clear showing a direct correlation between the two measures.[17] But these data are misleading because of the distortions of commoditization. Prosperity in this formulation is measured in dollars of per capita GNP, so what is being measured is the total amount of goods and services moving in the economy, which of course, requires vast amounts of energy for production and distribution. The very definition of prosperity is directly tied to energy use through commoditization. If it is possible to break the hold commoditization has over our economies, which I believe it is, than the hard link between well-being and energy use can be overcome to a considerable extent. In fact, improvements in a measure of well-being per unit of per capita fuel consumption would be an excellent indicator of progress in the decommoditization of an economy. It would also provide an excellent measure of the capability of a society to meet the needs of its people without damaging the natural world within which they live.

In 1985 Jose Goldemberg reported on the results of his analysis of the relationship between per capita energy consumption and the measure known as the Physical Quality of Life Index. The PQLI combines three measures that often are used as a way of quantiying the level of well-being in a country: infant mortality rate, life expectancy, and literacy. According to Goldemberg, "When the PQLI is plotted against per-capita energy use (commercial plus non-commercial) for a large number of countries, it is found that on average a PQLI of about 90 (a value typical of industrialized countries) is reached for per-capita energy use rates of 1.0–1.2 kW, and that further increases in energy use cause only a very marginal further increases in PQLI."[18] There was considerable variation in the data, and some countries reached a PQLI of 80 with only 0.5 kW per capita, while others achieved 90 with 1 kW.

How would technology evolve in a society not distorted by commoditization forces, a society that truly took account of the reality of entropy and was interested in energy efficiency? Design would begin with two questions: first, how can the need for transportation, food production, clothing, etc. be met using the least amount of energy and materials and second, how can the natural flows and cycles associated with the landscape and climate of a particular place be utilized so that no energy is wasted fighting against or unnecessarily altering the patterns of natural flow. Under the rule of commoditization these questions are rarely even asked. The answer to the question of whether the hard link between prosperity and energy use can ever be broken, depends entirely on whether we can get ourselves out of the commoditization trap and turn our attention to making real progress in designing, engineering and living with considerably reduced flows of energy and materials.

Once we shift economic goals from growth in the production of commercial goods and services to improving quality of life, the concept of efficiency takes on dramatic new meaning. E. F. Schumacher, evoking what he called a Buddhist economics, described the flaw in the standard notion of efficiency in his classic book *Small is Beautiful*. The modern economist, according to Schumacher, "is used to measuring the 'standard of living' by the amount of annual consumption, assuming all the time that a man who consumes more is 'better off' than a man who consumes less. A Buddhist economist would consider this approach excessively irrational: since consumption is merely a means to human well-being, the aim should be to obtain the maximum well-being with the minimum of consumption."[19]

The problem with Schumacher's and similar analyses is that they consider consumerism to be a personal, individual weakness rather than a pattern of behavior that is systematically reinforced by the structure of society through commoditization. Of course, greed and superfluous consumption can be overcome by individual decision, but it would be far easier for many more people to make that decision if society intentionally reinforced and rewarded frugal behavior and community-building activities. The policies that follow from the spiritual criticism of consumerism are the policies of personal change. Although conversion can powerfully influence behavior, the force of commoditization is systemic and powerful and largely independent of individual beliefs. That is why the economies of nominally Buddhist nations are as subject to commoditization as the most materialist, secular nations. Only through policies that address commoditization by rewarding thrift and penalizing waste can energy conservation be systematically internalized in the economy.

The technology of real energy conservation consists of goods and processes with inherently low commodity potential and commoditization has steered development away from them, including:

- An organized, well-supported infrastructure for product sharing, leasing, repair, and maintenance
- Improvements in reusability and recyclability of all goods
- Design and production in cooperation with natural flows of material and energy rather than against them
- Flexible design encouraging multiple uses for specific contexts
- Production designed for durability and simplicity of use

As we will see more fully below, these approaches suffer from an R&D famine precisely because each involves skills and practices that have inherently low commodity potential despite enormous social and environmental value.

7.2.3 *Efficiency is enhanced by working with natural flows and processes rather than against them: the principles of appropriate technology and ecosystem thinking*

Implication: Technology should be designed to work with rather than against natural flows of energy and materials.

The only sensible way to increase prosperity while decreasing energy and material usage is to design and produce *everything* with conservation and efficiency in mind from the beginning. The skills and knowledge to do this are widely available, and terribly underutilized. There has been nearly 30 years of independent, underfunded creative experimentation with what has been called "appropriate" or "alternative" technology.[20] Most such technology is based on working smart with the fewest and least environmentally disruptive of available materials derived from renewable sources, renewable energy, and designing *with* rather than *against* natural flows and local natural conditions. They require simple skills and tools familiar to any tinkerer or do-it-yourselfer and are relatively simple to build, maintain, and repair. The skills required to be an appropriate technologist are intimate, detailed knowledge of energy flow, experience with a variety of available materials, detailed knowledge of the intended use or service, basic skills of mechanical design and engineering, flexible problem-solving skills, and teamwork and leadership skills, since no one should have to cultivate all these skills alone.

When you utilize the energy of the wind in a sail or a windmill, you do not reduce or degrade the wind. When you design a home to soak up and store the warmth in its solar facing wall, you do not degrade the sunlight. When you work with organisms and compost to build soil you direct and enhance natural processes. When you utilize the energy of falling water to turn a mill you do not degrade or reduce the water. If you use wood at a rate no greater than it can grow you sustain the resource. Working with natural flows saves energy; working counter to natural flows expends energy. Real-life solutions to the problems of energy and materials conservation will include both hard technologies based on fossil fuels and soft technologies that utilize sources of renewable energy. There are no formulas for appropriate technology: that is its strength and its difficulty.

Serious conservation of energy requires a careful study of natural flows and possibilities of benefiting from their work: the flow of sunlight, the flow of water, the flow of wind, the flow of minerals, and the flow of nutrients. This is the heart of appropriate technology: observation, understanding, design, testing, and refinement through practice. Appropriate technology rests on the tinkerer's art. It always and profoundly resists standardization, packaging, and many of the other features of commodities. That is the biggest problem for appropriate technology in an economy subject to heavy commoditization pressures. This is why the practice of appropriate technology has remained largely underdeveloped despite centuries of slow and careful research and development by indigenous engineers everywhere in the world. The possibilities for appropriate technology are enormous and virtually untapped compared with, for example, the possibilities of packaging and transport.

The potential has been demonstrated by countless examples, none more remarkably than in the intentional community of Gaviotas in Columbia. With limited resources and geographic isolation, the innovators of Gaviotas have invented numerous small-scale systems that work with low-cost materials and minimal energy input. They have relied on continuous experimentation and the knowledge they gain from close observation of the ecosystem they occupy. As Alan Weisman describes in his book on Gaviotas:

> In 1971, a group of Columbian visionaries and techni-
> cians, reasoning that surging populations must some-
> day learn to inhabit even the world's harshest regions,
> decided to prove they could thrive in one of the most
> brutal environments on earth: their country's barren,
> rain-leached eastern savannas…. Sixteen hours from
> the nearest major city, they invented windmills light
> enough to convert mild tropical breezes into energy;
> solar collectors that work in the rain, soil-free systems
> to raise edible and medicinal crops, solar "kettles" to
> sterilize drinking water, and ultra-efficient pumps to
> tap deep aquifers — pumps so easy to operate, they're
> hooked up to children's seesaws.[21]

Not only did they succeed over more than 25 years to build a model sustainable community, but "in the shelter of millions of Caribbean pines, which the Gaviotans planted as a renewable crop, an unexpected marvel has occurred: the regeneration of an ancient native rain forest."[22] Through inge-nuity, experimentation, hard work, and respect for basic ecological princi-ples, the Gaviotans appear to have succeeded in building a human settlement that not only protects the natural environment, but enhances it.

The greater the ecological flow or force, the greater the amount of energy required to contain or disrupt it. For example, trying to contain a major river within its banks and insuring property owners against flood damage can be extremely energy inefficient compared with avoiding per-manent settlement of the flood plain to begin with. The experiences with flood control in the U.S. is a good example. While U.S. federal flood control expenditures increased dramatically each decade from the 1920s to the 1960s, so did the amount of flood damage. Flood control efforts succeeded in protecting property from the *average* flood, but when storms and flooding were extreme the resulting damage was far greater than had been the case prior to the institution of the flood control programs. The flood control program and subsidies for property insurance had encour-aged construction and development in many parts of the floodplain that had once been frequently inundated and therefore left undeveloped. Now when flooding does occur, there is far more real estate in its path and far more damage as a result. Furthermore, where in the past a flood might typically spread out wide and relatively shallow across the plain, now with the river confined to channels by levees, the floodwaters reached new and dangerous heights, causing floods in unleveed tributaries that would not have flooded before. Some rivers, particularly with large sed-iment loads, regularly shift channels as sediment build-up changes the patterns and gradients of elevation. Rigid flood control structures prevent channel-shifting, which means sediments continue to build up and dangerously raise water levels in the channel. Sediment management then also becomes an energy-intensive part of flood control to replenish and/or protect downstream deltas.[23]

Trying to stabilize anything dynamic requires energy, sometimes huge amounts of it. Humans routinely intervene in natural systems to make them more stable and predictable, and each intervention exacts costs in terms of energy demands and unintended consequences. We are in a tragic situation in that we must attempt to control the forces of nature for the comfort and safety of our loved ones and ourselves, but when we do so we must pay in ways that are sometimes as costly and damaging in the long run as the hazard we first attempted to avoid. The solution, if there is one, lies in a deep and detailed awareness and understanding of natural flows and processes so that instead of battling flows and stabilizing natural variability, we can accommodate and adapt ourselves and our technologies according to our best current understanding of ecosystem processes. When this is done at the relatively small scale of the individual or village, it is called "appropriate technology"; on the larger scale of a bioregion or ecosystem it is called "the ecosystem approach to resource management and economic development." The most important element for success with these approaches is a deep and detailed understanding of the properties of a given ecosystem, being able to design interventions that fit unobtrusively into that system, and moderate the impacts of human activities and the human economy on the system as a whole. The problem is, once again, that understanding systems requires a form of science that is inherently multidisciplinary, difficult, and time-consuming while resulting in knowledge that has little if any commodity potential. As a result, this science is starved for the kind of research and development it desperately needs.

Like the shifting channels of a river, most natural ecosystems are dynamic. Variability and change is a constant feature of nature. The natural history of most areas results in a patchwork rather than homogenous landscapes. Different parts of the system are subject to various cycles at different rates that affect key aspects of growth and development. Many of the features of an ecosystem exist precisely because it experiences periodic events of destruction by fire, flood, drought, disease, insect infestation, or other cataclysmic events. In the long natural history of many ecosystems, the native species have evolved mechanisms to survive the periods of high stress, and many require such outbursts of creative destruction in order to reproduce and survive. When this variability — be it fire, flood, infestation, or other natural "disaster" — is reduced to accommodate the human need to impose order and stability, then the condition under which the ecosystem had evolved is fundamentally altered, and gradually the ecosystem itself is transformed and the plants and animals depending on it are lost. Human enterprise often attempts to create homogenous landscapes — be they vast fields of corn or vast stretches of highway, or placid, easy-riding rivers — out of what was once patchwork. As this happens more and more throughout the world, the diversity of nature, based as it is on the diversity and variability of ecological conditions, gradually becomes homogenized.

C. S. (Bud) Holling studied 23 examples of managed ecosystems from different parts of the world.[24] In each case, the effort to suppress variability and to maximize a single target, socially beneficial variable led inexorably to decline and in some cases destruction of the managed ecosystem. Natural cycles affect the reliability of harvest: for example, periodic outbursts of spruce budworm in the eastern North American spruce/fir forests or outbreaks of fire in the western U.S. Sierra Nevada forests greatly reduce timber harvest; widely varying numbers of salmon in western U.S. streams lead to boom and bust cycles for the fishermen; varying density and species mix of rangelands lead to rising and falling beef production, and so on. Holling showed that in each of these cases resource managers attempted to stabilize natural variability. The budworm are attacked by insecticides, the fires suppressed, the grasses maintained with modern rangeland practices, the salmon stocked at varying rates to keep harvest numbers steady. What happened is that, without the variability, instead of many patches of varying ages and stages of impact or recovery, the ecosystem became more homogenous. Then, for example in the forest, when an outbreak of insects or fire did occur, which it inevitably did no matter how successful the suppression, it was far more destructive. Likewise, management practices turned rangelands into beds of highly productive but drought-sensitive grasses. Under drought conditions the range turned into a virtual desert with a few shrubs.

It is certainly rational for producers to attempt to minimize sudden fluctuations in yield or output. Predictable, preferably steady supplies of output are one of the defining characteristics of a successful commodity. Harvesting natural resources in a highly commoditized economy requires massive investment in equipment and labor. Predictable yields are needed to satisfy the demands of the investors. Without ecologically based policies to counteract the effects of commoditization, the logic of exploitation is inevitable and inevitably destructive. For the rational market-driven actor, there are really only two options: sustainable yield, which implies suppressing the variability of the commodity being produced, or worse, massive one-time harvesting as in clear-cutting rainforests and placing the one-time income into other investments with greater or more steady economic yields. The logic of commodity production is powerful. More ecologically based resource management alternatives inevitably have lower commodity potential because they require less capital and more labor; they require intense personal knowledge of the underlying ecosystem and rely on the existence of a mixed local economy where local timber suppliers are in active social relationship with timber consumers. For ecosystem management to be successful, a balance must be struck between commodity production and conservation of noncommodity values. That requires a coordinated system of economic and social policies to counter commoditization pressures and promote ecologically informed, ecosystem-based management.

As a guide for making judgments and determining right action, the most important principles of the ecosystem approach are as follow:

1. Our notion of community (and our responsibilities toward it) should include soils, water, plants and animals, and the system features that emerge from their interrelationships, or collectively: *the ecosystem* including the people who live on it and are shaped by it.
2. We should design management strategies for not just any ecosystem, but for specific ecosystems each with its own physical and biochemical construction, energy flows, and history; what is sometimes called "site-based management."
3. We should recognize that the natural history of ecosystems ultimately determines the sustainable productivity of any natural resource.

Governments are increasingly giving expression to ecosystem concepts such as the following example from Canada's Green Plan:

> We live in a complex and integrated environment. All creatures, including humans, interact with and depend on each other. They all draw on the materials and energy of the physical environment to obtain food and recycle wastes. They all affect each other's behavior. In the past, responses to environmental problems paid little attention to these inter-relationships. Today, the increasing number and complexity of environmental issues demand that we adopt a more integrated approach.[25]

Such pronouncements can be found in countless documents in countries all over the world. They ring hollow as long as the governments remain locked into a system of continuing economic growth based on commoditization. Economic and social reforms such as those discussed in Chapter 8 are necessary before ecosystem approaches to management and appropriate technology can become predominant in the economy.

Serious energy and material conservation involves not only designing tools but also communities, workplaces, transportation systems, and schools with ecological principles in mind. The most effective energy and materials conservation efforts involve integration and multidisciplinary cooperation to accomplish rational, efficient design: neighborhoods and downtowns built for walking and biking, bikes integrated with public transport, schools and factories integrated with transportation systems, heating and lighting integrated with landscape design, economic development integrated with land use planning and environmental protection. Such levels of integration require ecologically informed, broad participatory planning far beyond the capacity of our present governance systems.

Greater integration of production and consumption is also the key to reducing the incredible amounts of waste produced by modern economies. In nature, everything depends on cycles and loops. The waste from one activity becomes the source of energy (food) for the next activity. Considerable energy and materials efficiency could be gained by designing our systems of production and consumption so that every product or by-product of a process becomes an input for another process.

The trends driven by commoditization, however, go in just the opposite direction: toward easy transport, quick consumption, and ready discard. Most commodities follow a one-way, linear path from producer to consumer to the landfill or incinerator. To transform this waste-making machine into an economic system designed for efficient use of materials and energy requires that producers take responsibility for what is sometimes called the complete life cycle (cradle-to-grave) of their products. This is also referred to as *extended producer responsibility*.[26] With such responsibility would come the incentive to design long-lived products with consideration of their end use and/or reuse. Extended producer responsibility also would require integrating consumers and producers into closely aligned systems of materials management in which producers and consumer relationships continue throughout the use of the product. Leasing would become a viable option for most major appliances and things like carpeting. After the lease period ended, the producer would take the product back for repair or recovery of materials.[27] In effect, markets for major goods would shift to markets for the services which the goods provide.

Energy efficiency and appropriate technology, like everything else, is subject to commoditization pressures. There is a market for energy-efficient products, a market that would surely grow with increased energy prices. But such efficiencies, in and of themselves, will not decrease energy use if the dollars saved are merely transferred to the purchase of other commodities. Likewise, a market exists for the tools and skills of appropriate technology, and there is many a do-it-yourselfer who, as a leisure time activity, is retrofitting her home. Middle-class Americans and Europeans are experimenting with many forms of eco-villages and super energy efficient homes, which by themselves simply become a housing specialty market. The wholesale adoption of ecological technology and ecosystem approaches will require real change in the system of economic incentives that presently drive commoditization.

Fundamental to the successful adoption of ecologically sustainable ways of life is a significant shift in perspective by enough people so that the values of simplicity, grace, and connection with the natural world become predominant in society. Many observers and activists place their hope in environmental education and the widespread adoption of environmental ethics. Although both are extremely important, without change in the system of economic incentives, environmental awareness and the fostering of feelings of deep connection with the Earth will remain specialized areas of interest for a minority of self-selected individuals. Feelings of concern for the health of the environment and connection with the natural world can then be appealed to for the marketing of commoditized eco-goods and services, such as eco-tourism and wilderness gear. Although a small but significant lobbying voice for environmental protection can be organized and politicized around these feelings, the majority of society will continue in its profligate and carefree pattern without real economic changes that address the problem of commoditization.

The prerequisite for living out an environmental ethic is awareness. One cannot care about what one does not know, and certainly one cannot protect what one does not understand. We cannot follow the ecological principle of designing with nature if we do not understand nature's designs. As the need for environmental awareness grows, the reality, driven by commoditization of the land and agriculture, is that fewer and fewer people are actively engaged in working on the land and developing an intense consciousness of nature. It is not that people are not spending more and more time in nature as spectators or in various forms of outdoor recreation, but such is a very different experience from slow and careful work of cultivation and husbandry, requiring attention and observation skills of another level than sightseeing. While the intellectual content of environmental ethics has grown with a great number of publications and courses dedicated to articulating and further elaborating the arguments for such ethics, the practical relevance of its ethical principles remains about as marginal as it was when Aldo Leopold wrote his famous prescription.[28] The irony of the evolution of environmental ethics is that as its intellectual content has grown, its existential content has probably declined. Our ability to act on that ethic in a meaningful way has been reduced correspondingly.

The child in third grade who learns, as Leopold hoped she would, ecological principles, can think of no real application of that knowledge other than to use two sides of a sheet of paper to spare a tree. Meanwhile, her world is a cacophony of manufactured materials, much of it produced in ecosystems thousands of miles away, over which she has no control. Leopold wrote that, *"perhaps the most serious obstacle impeding the evolution of a land ethic is the fact that our educational and economic system is headed away from, rather than toward, an intense consciousness of land."* Activities that foster an intense consciousness of land — working soil, shaping materials, gathering food — remain segregated in arts and crafts and field trips to zoos and nature centers. Despite the best intentions of environmental educators, the environment remains, in this context, another form of entertainment.

Commoditization creates an economic selection pressure that maximizes the attributes of commodities as described throughout this book. Nearly every analysis of how to reduce our economy's negative environmental impacts concludes with some version of recommending that we redesign our economy and its industrial and commercial tools in ways that mimic the cycles and flows of nature. But the very attributes we want to mimic in nature are incompatible in many ways with the attributes that make something a successful commodity. As commoditization operates over time, therefore, it becomes increasingly difficult to follow the guidance of the natural world even as we learn more and more about the ways nature works and the importance of following its dictates.

7.2.4 *Contradictory goals cannot be maximized at the same time and must be balanced: The principles of homeostasis and optimality*

Implication: Information indicating that the human economy is out of balance with nature must be received and processed and adjustments made to optimize the sometimes conflicting goals of prosperity and ecological integrity.

Life is full of contradictions that can never be fully reconciled; they can only balanced while we do our best to navigate some optimal middle ground, like Goldilocks choosing not too hot and not too cold, not too soft and not too hard, not too fast and not too slow. Every action, every creature, and every life operates best under optimal conditions that represent a balance between countless variables in complex patterns of relationships with each other. This is also true of nonliving matter as well. Each structure represents a temporary point of stability between several chemical, physical, and biological forces acting upon matter in often conflicting ways. Within these points of balance life evolves in wildly diverse ways that take advantage of flows of energy and nutrients temporarily made reliable and semipredictable by the improbable structure. Structure is always a highly improbable balance temporarily achieved between counteracting forces. If any one of these forces happens to be maximized to the loss of the others, then the structure is destroyed. Although no single force is ever maximized in any structure or life or pattern of any kind, each is being optimized in relation to each other. This optimum is always temporary, because everything is changing over time and as each variable changes all the others react. The miracle of life is how much stability actually exists over time. The way life maintains stability is what we call *homeostasis*, the ability or tendency of an organism to maintain internal equilibrium of temperature, fluid content, and countless other variables by the regulation of its own internal physiological processes and by automatically adjusting to changes that occur in the external environment. The most important aspects of successful homeostasis are:

- The capacity to continuously gather, receive, and evaluate information coming in from the environment.
- The ability to respond appropriately to the information by adjusting internal conditions in ways that maintain favorable balance.
- An external environment governed by its own homeostatic mechanisms that maintain conditions roughly within the range of extremes of the particular species' evolutionary experience.

Homeostatic mechanisms exist at the level of the cell, the organ, and the organism. Organisms interact in complex relationships with each other and their shared environment and maintain some stability over time in recognizable ecosystems such as lakes, forests, deserts, fields, and prairies, each with its own characteristic group of organisms that have evolved within the conditions each creates for and with the others. The homeostatic processes of ecosystems maintain conditions within ranges of extremes, which thus enhance the long-term survivability of the component species of organisms that utilize the resources of a given ecosystem. The Earth as a whole likewise maintains stability through complicated feedback mechanisms that constitute a form of homeostasis at the planetary level. Balance is achieved among many counteracting forces and is sustained as long as no single variable is maximized at the expense of the others.

Tradeoffs in nature are similar to the experiences we have in our daily lives. In nature much of the tradeoffs have to do with efficiently obtaining, allocating, and using life's primary scarce resource, fuel (food), to survive and then reproduce. The energy that goes into mating may be lost to feeding, while that which powers transportation may not be available for mating. What can be done quickly often cannot be done efficiently. What can be done efficiently often cannot be done on time. The economy of life's energies can be balanced in countless ways, and the millions of species of plants and animals each represent another set of balanced solutions to the problem of survival and reproduction.

The goal of economic development is often to maximize single variables in complex ecological relationships, variables that are favorable in some ways to humans. Thus agriculture is the attempt to maximize the yield of harvestable nutrients by manipulating the environment in such a way that more of its energy goes into the production of these nutrients. Manipulating the flows of energy and nutrients in the environment to maximize the production of beneficial products is the very essence of the human economy. But the more we attempt to maximize certain attributes over others, the more we manipulate the natural environment, the more energy is required to sustain production at this new imbalance. Intelligent management of the environment, based on conserving materials and energy, would require minimizing the amount of overt manipulation of the environment that is done to maximize economic production. The goal of the new economy should be to achieve optimal human welfare with the minimal of environmental disruption.

Homeostasis is very much about information processing. Human intelligence is an advanced homeostatic tool. It not only takes in information and compares it with some decision rules programmed into it by evolutionary experience, but humans can make judgments apart from rules, forecast, learn from others' experiences, and think about the meaning and method of intelligence itself. Intelligence allows us to continually evaluate our environment and change our behavior accordingly. We also work cooperatively with each other in complex social systems designed, at least in part, to optimize the well-being of all its members. This is the most hopeful fact of our current environmental predicament. Humans have the real capacity to learn from the environment and make adjustments. If the human economy is organized in ways that are threatening to disrupt the Earth's homeostasis and balance, then humans have the capacity to recognize this fact and adjust the economy accordingly.

We simply cannot maximize production or continually manipulate the environment to maximize material satisfactions. The best we can do is to find a reasonable balance in which we optimize material well-being within self-imposed economic constraints designed to protect the balance and stability of the Earth's life support systems.

The environmental crises, both current and to come, demand that we manage our impacts on ecosystems in ways consistent with our understanding of ecological principles. There is no way to avoid the need for effective economic policy informed by ecological principles. Commoditization will often create positive feedbacks where negative ones are what is needed. For example, when a resource begins to become depleted (fish, for example), economic signals cause the price to rise. As a result of rising prices, harvesters (fishermen) are encouraged to invest more to improve yield to obtain more of the product at the higher prices. The better equipped harvester just hastens the collapse of the resource (fishery). This is played out over and over again with scarce resources. Optimistic free-market economists argue that higher prices encourage the development of substitutions for scarce resources (plastic for wood, synthetic fibers for wool, rayon for rubber, aquaculture for fishing) and decrease the pressure on the resource. Although this has been true on occasion, each of these substitutions involves synthetic substitutes that not only increases the amount of fossil fuel energy embodied in each unit of product but also often create vast new quantities of waste, many of which involve new synthetic chemicals that do not break down in the environment. The end result of these substitutions driven by commoditization is an increase in energy use and pollution. The logic of a system (the economy) and a discipline (economics) that cannot value Nature, only its economic functions, is an illogic that distorts our thinking and sickens our Earth.

7.2.5 *Scale and level of organization matter: the principle of cooperative hierarchical organization*

Implications: Economic policy decisions should simultaneously consider effects at the level of the individual, the level of the economic system, and the level of the global ecological system.

For any structured activity in nature there is an optimal size. When designing systems to serve human welfare with the least environmental impact, the appropriate scale of the system is an important consideration. If the purpose is to maximize the production of marketable goods and services the infrastructure may be designed to operate at a very different scale than if the goal is to optimize human welfare in conjunction with minimizing environmental impact. In most but not necessarily all cases, the scale of industry designed to maximize commodity production will be much larger than the scale for energy-efficient human welfare optimization.

Ecologists often think of the world in terms of hierarchies of scale: mole-cules–cells–organs–organisms–ecosystems–biomes–biosphere–solar sys-tem–universe. A forest has many energy and material flows that exist at different scales and rates of exchange and operate at a particular scale of organization: the cellular level, the level of leaf litter and soil, the level of the whole tree, the level of the stand, the level of the forest ecosystem, which interacts with and is affected by surrounding conditions up to the level of the biosphere. Depending on what you want to know about how the forest functions and what you want the forest to provide, you may ask questions focused at completely different scales and sets of exchanges. For example, you might ask how does a plant capture energy from the sun? or How is forest soil produced and reproduced? or Who eats whom among the creatures of the forest? or What conditions are optimal for the production of hardwood for mills? What conditions are optimal for the production of pulp for paper? What conditions are needed for the forest ecosystem to be sustained indefinitely into the future? It's important when seeking answers to questions about any multilayered system to target one's observations and measures at the appropriate scale.

Our human institutions have been largely shaped by the needs of com-moditization. Our capacity to act collectively has also been similarly distorted by commoditization. The boundaries of local jurisdictions are usually artifacts of recent history. They either served the convenience of colonial administra-tion, trade routes, or the land survey that preceded the sale and enclosure of once open or public lands. Throughout history, governance and regulatory authority has shifted to different scales along with the expansion of trade and the interest of establishing common standards for commerce and the enforce-ment of contracts throughout an economically integrated area. At the end of the twentieth century and the beginning of the twenty-first, we are witnessing the growing integration of the commoditized economy at the international and global scale and an increasingly powerful set of global institutions whose task it is to encourage commoditization worldwide and facilitate the further integration of the global economy. We now have a hierarchical pattern of institutions that encourages the flow of commerce at the local, subnational, national, international, and global scales. At present there are only weak insti-tutions whose interests lie in the welfare of people and the protection of ecosystems from the effects of all this increased commoditization.

If we are to have the ability to effectively defend the environment, promote conservation, and provide countervailing pressures to commoditization, we will also require the ability to act collectively at the scale most consistent with ecosystem levels of concern, be it local, regional, or global. As long as commoditization drives human activities, the scale of human intervention will remain at the level at which commodities function. If we want to manage processes as well as products, systems as well as commodity movement, then we will need new structures of governance that create the possibility for people to act collectively at the appropriate level. Some possible suggestions for how this might be organized are discussed in the next chapter.

7.3   Conclusion

The field of ecology has taken great strides in demonstrating its relevance to economics, so much so that economists ignore the conceptual frameworks and warnings of ecologists only at great peril to their profession and to the world. Indeed, ecology is more even than an academic counterpart to economics, but should better be understood as a metadiscipline without which economic analysis is fundamentally incomplete and inadequate to make long-term projections of economic development. As the analysis in this chapter has sought to show, we cannot understand our present economic condition or make rationally justifiable decisions about our economic future without taking into consideration ecological principles that simultaneously make opportunities available to us and constrain our options. If we intelligently apply ecological principles to our economic analysis and decision making, than we will have many more opportunities than constraints. If we wait to analyze our condition in ecological–economic terms until ecosystems are stressed beyond repair, then surely we will be faced with overwhelming constraints and few opportunities.

The five ecological principles described in this chapter do not exhaust the economically relevant insights of ecology. They do begin to demonstrate just how far askew commoditization has taken the human economy. They also hint at the direction we need to take in order to bring the human enterprise in line with what we know about the way the world works. The following chapter discusses a general program of policies that might just do the job.

Notes

1. Daly, H.E., *Beyond Growth: The Economics of Sustainable Development,* Beacon Press, Boston, 1996. Douthwaite, R., *The Growth Illusion,* Council Oak Publishing, Tulsa, OK, 1993; see also Goodland, R., Daly, H.H. El Serafy, S., and Von Droste, B. (Eds.), *Environmentally Sustainable Economic Development: Building on Bruntland,* UNESCO, Paris, 1991.
2. Erekson, O.H., Loucks, O., and Strafford, N., The context of sustainability, in *Sustainability Perspectives in Business and Resources,* Loucks, O., Erekson, H., Bol, J.W., Grant, N., Gorman, R., Johnson, P., and Krehbiel, T. (Eds.), CRC/St. Lucie Press, Delray Beach, FL, 1998.
3. Daly, H.E., From empty world economics to full world economics: recognizing an historic turning point in economic development, in Goodland, R., Daly, H.H., El Serafy, S., and Von Droste, B. (Eds.), *Environmentally Sustainable Economic Development: Building on Bruntland,* UNESCO, Paris, 1991.
4. Rees, W.E., Patch disturbance, dissipative structure and ecological integrity: a "second law" synthesis, in *Global Ecological Integrity,* Pimentel, D. (Ed.), Island Press, Washington, D.C. (in press).

5. Folke, C., Jansson, A., Larson, J., and Costanza, R., Ecosystem appropriation by cities, *Ambio*, 26, 167, 1997; IIED, Citizen Action to Lighten Britain's Ecological Footprint, Report prepared by the International Institute for Environment and Development for the U.K. Department of the Environment, International Institute for Environment and Development, London, 1995; Rees, W.E., and Wackernagel, M., Ecological footprints and and appropriated carrying capacity: measuring the natural capital requirements of the human economy, in *Investing in Natural Capital: The Ecological Economics Approach to Sustainability*, Jansson, A.M., Hammer, M., Folke, C., and Costanza, R. (Eds.), Island Press, Washington, D.C., 1994; and Wackernagel, M., Onisto, L., Callejas Linares, A., Lopez Falfan, I.S., Mendez Garcia, J., Suarez Guerrero, A.I., Suarez Guerrero, Ma. G., *Ecological Footprints of Nations, Report of the Centre for Sustainability Studies*, Universidad Anahuac de Xalapa, Mexico and The Earth Council, Costa Rica, 1997.

6. IIED (1995).

7. Rees, "Patch disturbance" (in press).

8. Vitousek, P.M., Ehrlich, P.R., Ehrlich, A.H., and Matson, P.A., Human appropriation of the products of photosynthesis, *Bioscience*, 36(6), 368, 1986.

9. Vitousek, P.M., Mooney, H., Lubchenko, J. and Melillo, J., Human domination of Earth's ecosystems, *Science*, 277(25), 494, July 1997.

10. Rees (in press); Peet, J., *Energy and the Ecological Economics of Sustainability*, Island Press, Washington, D.C., 1992; Ehrlich, P.R., Ehrlich, A.H., and Holdren, J.P., Availability, entropy, and the laws of thermodynamics, in *Valuing the Earth: Economics, Ecology, Ethics*, Daly, H.E. and Townsend, K.N. (Eds.), MIT Press, Cambridge, MA, 1993; Georgescu-Roegan, N., *The Entropy Law and the Economic Process*, Harvard University Press, Cambridge, MA, 1971; and Daly, H.E., On Nicholas Georgescu-Roegen's contributions to Economics, *Ecological Economics*, 13, 1995, 149–154.

11. Ehrlich, Ehrlich, and Holdren (1993), p. 71.

12. Hall, C.A.S., Cleveland, C.J., and Kaufmann, R., *Energy and Resource Quality: The Ecology of the Economic Process*, University Press of Colorado, Niwot, CO, 1992, p. 50.

13. Reddy, A.K.N. and Goldemberg, J., Energy for the developing world, *Scientific American*, September 1990, p. 112.

14. Ko, J.-Y., Hall, C.A.S., and Lopez Lemus, L.G., Resource use rates and efficiency as indicators of regional sustainability: an examination of five countries, *Environmental Monitoring and Assessment*, 51, 571, 1998.

15. Ko et al. (1998).

16. Schmidt-Bleek, F., MIPS — a universal ecological measurement, *Fresenius Ecological Bulletin*, 1, 306, 1992; also in *Fresenius Ecological Bulletin*, 2, 8, 1993, a special edition on the Material Inputs per Unit of Service [MIPS] project of the Wuppertal Institut für Klima, Umwelt, und Energie, Wuppertal, Germany, 1993; Young, J. and Sachs, A., *The Next Efficiency Revolution: Creating a Sustainable Materials Economy, Worldwatch Paper 121*, Worldwatch Institute, Washington, D.C., 1994; Rees, W.E., Reducing the ecological footprint of consumption, *The Business of Consumption: Environmental Ethics and the Global Economy*, Westra, L. and Werhane, P. (Eds.), Rowman and Littlefield, Lanham, MD, 1998.

17. Hall, Cleveland, and Kaufmann (1992).

18. Goldemberg, J., Johansson, T.B., Reddy, A.K.N., and Williams, R.H., Basic needs and much more with one kilowatt per capita, *Ambio*, 14(4–5), 190, 1985, p. 192.

19. Schumacher, E. F., *Small is Beautiful,* Harper and Row, New York, 1973, p. 57.
20. For examples, see Dickson, D., *Alternative Technology and the Politics of Technical Change,* William Collins & Sons, Glasgow, 1974, and Nadis, S. and MacKenzie, J., *Car Trouble,* Beacon Press, Boston, 1993.
21. Weisman, A., *Gaviotas: A Village to Reinvent the World,* Chelsea Green, White River Junction, VT, 1998.
22. Weisman (1998).
23. Hall, Cleveland, and Kaufmann (1992), p. 524.
24. Holling, C.S., What barriers? what bridges? in *Barriers and Bridges to the Renewal of Ecosystems and Institutions,* Gunderson, L.H., Holling, C.S., and Light, S. (Eds.), Columbia University Press, New York, 1995.
25. Environment Canada, *Canada's Green Plan for a Healthy Environment,* Government of Canada, Ottawa, 1990.
26. U.S. Environmental Protection Agency, Extended Product Responsibility: A New Principle for Product-Oriented Pollution Prevention, June 1997, Website report at http://www.epa.gov/epaoswer/non-hw/reduce/epr/epr.htm.
27. Interface Flooring Systems, Inc. has pioneered product take-back in the U.S. See their Website at http://216.1.140.49/us/company/about/.
28. Leopold wrote, "A thing is right when it tends to preserve the integrity, stability and beauty of the biotic community. It is wrong when it tends otherwise." Leopold, A., *Sand County Almanac,* Oxford University Press, New York, 1968, pp. 224–225.

chapter eight

Toward a coordinated decommoditization strategy

Contents

8.1   Introduction

This book has sought to demonstrate that over the course of centuries first the Western economy and then the global economy developed a structural bias in favor of commoditization. In spite of the material comforts and cultural benefits that commoditization has brought, it also has many profoundly negative consequences for humankind. These include the systematic impoverishment of the nonmarket aspects of social life, including the nurturing of families, communities, and other human relationships, and the care for the Earth. It is now time to consider some corrective actions that we can take. Before examining our options, however, it will be helpful to review the logic of the argument that has been made in these pages:

- The human economy and human societies are wholly dependent subsystems of the natural world and global ecological relationships.
- The material and energy used and harmful byproducts released by the economy are threatening to disrupt the natural world and seriously threaten the health and well-being of the Earth, many of its creatures, including humans, and the diversity of ecological systems that have evolved here.
- The solution to this crisis is to significantly reduce the impact of human economic activities on the Earth's environment.
- There are three ways to decrease the burden humans place on the Earth's ecosystems, the first two of which we would hope to avoid: (1) reduce the human population through mass death and shortened life spans; (2) reduce economic prospects, increase poverty, and eliminate the chance for prosperity for most of the world's people; or (3) reform the human economy so that it produces the prerequisites for a good life for all while limiting its physical impact to levels compatible with a healthy environment.
- The technology and social changes required to accomplish option 3 above are well understood but grossly underdeveloped.
- This underdevelopment is a direct result of commoditization, which operates as a systematic selection pressure favoring those goods and services with characteristics that make them most fit as commodities for exchange.
- Any selection pressure operating in a self-organized system such as the human economy acts over time to gradually expand the number of entities that carry preferred characteristics, thus crowding out and eliminating those entities that lack the preferred characteristics.
- The technology and social institutions necessary for sustainable development will promote goods and services with the following characteristics, which lack commodity potential:
  - They are customized to specific geographic, ecological, and cultural conditions.

- They require craftsmanship and stewardship grounded in deep knowledge of the specific conditions of end-use.
- They are built on relationships of caring, sharing, and collective effort among people.
- They are based on respect and understanding of the natural world and the principles of ecology and ecosystem dynamics.
- They are simple, thrifty, and efficient.
- They are built to last, to be easily repaired and upgraded or recycled into other goods or into the cycles of the earth.

- Since these characteristics are exactly those that are selected out by commoditization, if sustainable development is to be achieved, a decommoditization strategy of political intervention in the economy is essential to counteract the effects of commoditization.
- Public concerns about the state of the environment can be channeled toward economic reform if the public understands the ways in which the structure of the economic system presently undermines efforts to restore and protect environmental values and the possibilities for adjusting social rules and practices in ways that are consistent with ecological facts.
- The only way to limit market forces without diminishing freedom is if the capacity to regulate markets, raise resources for public use, and allocate public resources toward the common good is actively controlled by democratic institutions accountable to an engaged public.
- A "green" economy and a "green" democracy are best organized in a hierarchically nested series of democratically controlled jurisdictions that mirror the ecological organization of the planet.

Such dramatic political changes seem implausible, but they are realizable if we continue to raise the general public's awareness of the accelerating pace of environmental change and its causes, and if people continue to become aware of and experiment with alternative economic relationships. What is now a cause for concern among many people can then gradually become a powerful social movement that makes the desired changes possible, perhaps unavoidable. It is my position that without a fundamental change in direction, we will likely face an unthinkable future of mass impoverishment and suffering. We cannot afford to shut our eyes to the immensity of the problems we face, and there is no doubt, as the saying goes, that if we don't change direction we will end up where we're heading.

Human beings have an enormous capacity for action, creativity, and compassion. Although nothing can be predicted with certainty, we can be optimistic that as knowledge of both the problems and the solutions becomes widespread, we will begin to demand the economic and political changes needed for sustainable development. The most effective policies will be those that counteract the effects of commoditization and weaken its power over the direction of economic and social development. By focusing on commoditization, we can overcome one of the greatest obstacles to social change: the feeling of being overwhelmed by our many, seemingly disparate problems. Because commoditization distorts economic and social relations in so many areas, addressing its rule over the economy as a whole would have a broad, positive impact on all aspects of economic and social relations.

8.2   The policy wedges

This final chapter is about the policies that are needed to decommoditize the economy, and the social movement required to create the conditions in which these policies can be adopted and implemented. The proposed policies taken together constitute a coherent program for building an economy that is responsive to environmental and social needs rather than simply the interests of those who benefit most from the commoditized economy.

Commoditization as a force shaping economic and social relations is a natural outcome of the release of creative energy for personal wealth production. As a result, there is continuous improvement in the quality and availability of commercial goods and services, which greatly enhances the quality of human life wherever it is free to operate. The point is not to destroy the incentives for commercialization and marketing of goods and services, the point is to create enough countervailing pressure to end the domination of commoditization over all spheres of social development. What is needed is an equivalent release of creative energy in the direction of building and conserving natural and social wealth.

"Ecological footprint" and other analyses show a fairly wide variation in the amount of environmental impact per unit of population, even among industrialized countries.[1] Similarly, the Genuine Progress Indicator shows that in the industrialized countries the expected correlation between increasing GDP and improved quality of life has broken down since at least the 1970s (see Figure 2.3). Much of the consumption in affluent societies is above that needed for a secure and healthy life. These data suggest that the hard link between economic growth and public welfare is exaggerated and that there is considerable room to develop the noncommodity sectors of society.

To accomplish this development two things are necessary but perhaps not sufficient: first the capacity and willingness of society to invest in public noncommodity services that improve the quality of family and community life without necessarily increasing production, and second, the responsiveness of government to needs other than those of commercial and industrial interests. In other words strong, effective, and broadly representative governance.[2]

The challenge is daunting to say the least. As we have seen, the economic bias for commodities is structural and systemic. That means that environmental and social policies intended to stimulate sustainable development, no matter how well-intentioned, usually fail to counter the underlying pressures of commoditization. Such efforts are overwhelmed by unsustainable economic forces. Yet by understanding the systemic nature of the bias for commodities, we can design a suite of policy initiatives which produce countervailing economic pressures to minimize the bias. If decommoditization policies reinforce each other in a systemic way then they can become part of the economic structure within which producers and consumers make rational economic decisions. Such initiatives should have the effect of dramatically redirecting research, development, and investment toward the common good in a systemic, self-reinforcing way.

What we need are policies that improve quality of life without necessarily increasing the amount of energy and material throughput in the economy. There are three approaches, or policy wedges, that combined can reduce energy and material consumption while maintaining a high quality of life (Figure 8.1). These include

- *Improved efficiency* — getting more output per unit of input
- *Conservation* — conserving energy and materials whenever possible
- *Consumption-reduction strategies* — encouraging actions and behaviors that lead to voluntary simplicity and lower levels of individual consumption.

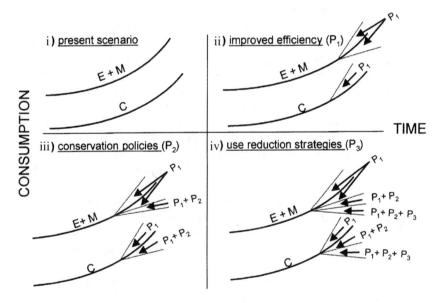

Figure 8.1   In the present scenario (i), increasing consumption (C) is tightly related to energy and material throughput (E + M) in the economy. (ii) In order to decrease throughput, economic activities must be made more energy- and material-efficient. Policies that stimulate improved efficiency alone can be expected to lead to lower prices and increased consumption, sometimes leading to increased throughput. (iii) Policies that encourage energy conservation (P_2), when combined with improved efficiency ($P_1 + P_2$) can begin to reduce consumption and E + M throughput. (iv) Further policies that encourage cooperation sharing and community provisioning, leasing, and extended producer responsibility and other use reduction strategies (P_3) can be combined with efficiency and conservation to significantly reduce energy and material throughput while maintaining high quality of life.

Although there is considerable overlap between these strategies, in general efficiency focuses on technical improvements in the production process, conservation focuses on the improvements in use and consumption, and reduction strategies focus on individual and social changes that encourage simpler, more environmentally aware patterns of consumption and use. For example, consider the effort to reduce the amount of fuel used to heat a home. Efficiency gains might come from technical improvements in the heating system, conservation gains would be made by insulating the home or adding south-facing windows, consumption reduction gains would be made by turning down the heat.

8.3   The role of participatory democracy

The most effective defense that a society has against the continued spread of commoditization and the primacy of private over public interests is a healthy civic life. Civic life refers to the participation of individuals in the political and cultural life of their communities and nation. It refers especially to such participation that is aimed at promoting the general welfare — actions meant to help others and not just oneself. Efforts to promote civic life locate the source of power in a society where it belongs, with its citizens.

The participation of citizens in their own governance is a manifestly practical ambition — practical in that self-governance is fundamentally a process of solving problems together. The solutions to problems that communities face, from crime to the lack of open urban spaces, require the participation of citizens, and yet that participation is often inhibited by rigid bureaucracies and the influence of corporate interests. Politics in which citizens are kept apart — in which there are few local forums for citizens to discuss and act on current issues — leaves important decision making with those who already have political and economic power, and for whom continuing the present course of economic development is most beneficial. In order to give local citizens more power over decisions and policies that affect them, they must have economic power. They must be able to determine the fundamental character of their communities, finding a balance between commoditized economic growth and public welfare.

Governance capacity is the ability to translate the collective will of the people as it is expressed through the institutions of civic life into coordinated effective action. There is simply no way to counter the distortions of commoditization without including the full environmental and social costs into the prices of goods and services as well as implementing policies that effectively shape the economic context within which private investment and R&D decisions are made.

Most of the arguments for sustainable development, ecosystem and adaptive management, appropriate technology, and environmentally friendly changes come down to an argument for intervention into economic decision making — intervention informed by ecological understanding. There is no way to intervene into the aggregation of private economic decisions collectively known as "the economy" without using the authority of government to act on behalf of the common good.

If we are going to expand the power and capacity of government, then we also must make government more accountable to an organized and active public, which is why the vitalization of civil society and democracy is so important. Without democratic institutions, increased governmental power becomes authoritarian. Authoritarian regimes are necessarily dependent on a ruling elite with centralized authority. The effect of authoritarian control over resources is similar to and more extreme even then the effects of commoditization. Authoritarian regimes thrive by intimidation and making citizens dependent on the central authority for their survival. Thus, the production and distribution of basic economic goods and services must be controlled by forces outside local communities.

Just as in a highly commoditized industrial economy, the goal of economic policy in authoritarian regimes is economic growth measured in GNP, because the wealth produced by such growth is readily transportable, centralizable, and assignable, and can be expropriated by the authorities whose power is thus enhanced. Noncommodity development in the forms of increased self-reliance, community and neighborhood empowerment, and collective capacity is as deeply threatening to authoritarian regimes (whether socialist or military) as it is irrelevant to capitalist-dominated industrial democracies. The collectivization, mechanization, and industrialization of agriculture under Stalin is a good example, as was the lack of attention to the environmental consequences of industrialization in the Soviet Union and Eastern Europe. A number of agricultural campaigns were launched under state socialism. These campaigns were centrally controlled, down to such details as when to sow seeds. Sowing had to be completed and reported to the authorities by a predetermined date regardless of local soil and climate conditions. Planting methods, such as whether to plant corn in rows or squares, were prescribed. The timing of grain harvest, cutting, and threshing were established by similar government dictates. The local indigenous knowledge of soil and climate was disregarded, and farming expertise soon atrophied.[3] The only outcome that mattered was whether production quotas were achieved. This is how commoditization can result even without the commercialization pressures of a free market. On paper, Communist regimes had strict environmental laws. In the day-to-day management of industry, however, output was nearly the sole indicator of performance. The noncommodity value of pollution prevention could not compete with the commodity value of production. Because the state was both the owner and regulator, preference was given to the objectives of primary concern to the authorities, which was economic output and growth — in spite of the fact that the government's policies ultimately undermined the capacity for the growth it sought.

Free market and authoritarian economies each are skewed toward commoditization and all its attendant environmental and social distortions. State-controlled economies have the propensity to become tyrannical and to stifle innovation and creativity and destroy local cultural diversity. Free market societies, while usually espousing the value of creativity and diversity, distort them by channeling them overwhelmingly toward commercial ends. What has often been presented as the only two political options — free-market capitalism or state-controlled socialism—each distort and limit human freedom and creativity and threaten the Earth.

If we are going to imagine and innovate our way out of the environmental crises of our times, we must be able to think beyond the confines of the false choices that are presented between two extremes. These dichotomies limit our thinking by presenting options that are equally unattractive. The choice is not between the tyranny of the market or tyranny of the state. Instead, there is an array of options for how humans can organize themselves and act collectively on behalf of the common good. The more we understand about the physical environment and the unintended consequences of our choices, the better chance we have of organizing our governments and our economies to minimize the negative effects of human activities. The choice, for example, is not between controlling nature or living passively as a victim of natural forces. Instead, there are countless ways to find a balance. The more we understand the forces of nature and the effects of human activities, the better we are able to make choices and remain flexible, capable of changing our choices as new information becomes available. Once again, as we have seen throughout this book, two extremes represent a false choice. The solution to the problems of commoditization is not to eliminate free markets but to use the collective human power expressed through government to alter the conditions and climate under which free markets operate.

We need effective democratic government that can intervene to deliberately and thoughtfully adjust economic system dynamics for the common good. One of the biggest obstacles is the way government itself becomes distorted by commoditization. As a result we are in a bind. We need government to counteract the commoditization effects of free markets, but government itself is affected by commoditization in ways that limit its effectiveness for acting on behalf of the public good. **This is why the first step in addressing the environmental crisis must be the emergence of a powerful social movement able to extract government from its capture by the forces of economic power.**

Such a movement could initiate the building blocks of new international, regional and subnational institutions with the capacity to effectively carry out policies at the local, national, regional, and global scales, and to encourage the expansion of civil society at all these levels.

The failure of modern societies to clearly identify the distorting effect of commoditization is a result in part of the institutional and ideological bias in which freedom is associated more closely with choice in the market than with self-government. Yet self-government is a primary democratic value, and economics should be subordinate to it. As important as markets are, it is time to rediscover the balance between market forces and politics wherein political judgments can be made that reflect a concern with the common good. The common good is first and foremost to be found in the commons — those aspects of life that are much less commoditizable because they involve substantive relationships among individuals and between individuals and their environment.

With the globalization of economic competition now taking place, the pressure to innovate has increased even further. It is to be expected that social uncertainty arising from insecure employment and rapid technological change coupled with steady global economic growth will further increase. Given the close association between economic growth, commoditization and ecological deterioration, the daunting challenge of reforming growth to minimize the negative effects on ecosystems must be faced right away.

8.4   *The powers of government*

Governments have six essential powers; they:

1. Create and support stable infrastructure for the provision of public goods and services, including the basic legal and organizational infrastructure of government itself.
2. Provide public goods and services.
3. Regulate activities, including commerce, that affect public welfare.
4. Collect taxes and expend public funds.
5. Protect human rights.
6. Provide for collective defense, public safety, and order.

These governmental powers, if used effectively as part of a coordinated decommoditization strategy, could counteract the effects of commoditization. But these powers themselves are subject to the forces of commoditization in the following ways:

1. *Government infrastructure* — There are two main aspects of the commoditization of public infrastructure:
    - In any given sector, infrastructure investments flow toward the most commoditizable options so that, for example, investments in building and maintaining roads and highways are far greater than investments in public transportation.
    - Elections are treated as a market exchange where candidates are sold to a public much like any other mass product.
2. *Public services* —
    - Whenever it is possible to profit from the provision of services, these services are commoditized and moved from the public to the private sector. As a result, private industry siphons off the simplest and most commoditizable tasks, leaving government with the most difficult and costly ones. As a result, governments always appear to be less efficient in delivering services than private industry.

- The pressures to reduce costs force governments to reduce labor costs through increasing labor productivity, with the result being the same commoditization substitution as in the larger economy: tools and products replace workers and relationships.
- Public services are preconceived as being the provision of services for the smooth functioning of the private economy. Education is transformed into worker training, public health into the medical industry, etc. Over time, public education, public hospitals, welfare and other social services become increasingly commoditized.

3. *Regulation* — The primary role of regulation becomes to encourage commoditized economic growth by minimizing uncertainty and maintaining favorable economic conditions, allowing corporations to plan and operate in a predictable and secure environment that permits profits over the long run. Everything from local zoning laws to international trade agreements are designed with the intention of encouraging economic growth in terms of increasing the volume of commodities. Environmental regulations focus on the most commoditizable functions of pollution control and treatment rather than prevention.

4. *Tax collection and public expenditures* — As a result of commoditization, governments gradually become major customers that pool the public's cash for use in purchasing goods and services.

5. *Protection of human rights* — Human rights become subsumed into the rights of producers and consumers. The expenditure of money is protected as speech and given free reign. The rights of property are treated as more sacred than the rights of the person.

6. *Defense and public safety* — As in other areas we have discussed, prevention, in this case of war and violence, ought to be the goal. But because of commoditization, peace-making rarely receives the same amount of R&D attention as weaponry. Similarly, it is easier to spend public resources on policing rather than community building and crime prevention. In times and places where public services deteriorate, the first public service to be privatized by the wealthy is often the police function as the market for private security forces grows.

8.5   Steps in a decommoditization strategy

8.5.1   Getting private money out of politics

As a result of the commoditization of governance, government becomes in substantial part an economic development agency whose success is measured by the ability to create and sustain conditions for economic growth. As we have shown, economic growth as typically measured is synonymous with commoditization. In the end, the effects of commoditization on government are to make government power itself a commodity that can be bought and sold by those wealthy enough to pay the price. Therefore, the necessary first step in revitalizing government as a force for the common good is to get private money out of politics as much as possible through public financing of elections and public access to mass communication for political debate.

8.5.2 Supporting LCP public services

Next, we need an active public to insist that government invest in and maintain support for those aspects of public welfare that have the least commodity potential. Every sector of the economy has a group of people interested in defending the least commoditizable aspects of that sector. These include advocates and supporters of

- Environmental conservation and protection, including pollution prevention and the protection of biological diversity
- The protection of small farmers
- Organic agriculture, including farmers and consumers
- Alternative health care, including providers and users
- Public health and health maintenance
- Public transportation, including providers and users
- Crafts and farmers' markets
- Public education, including teachers and parents
- The rights and sovereignty of indigenous people
- Cultural and ethnic diversity
- Prevention of crime and violence
- Care for the young and old
- Elimination of racism and other oppression
- Worker health and safety
- Family values
- Community building
- Full employment
- And many similar causes

These causes are all linked by the fact that the success of their efforts requires collective action against the effects of commoditization. This fact creates the conditions for building powerful *coalitions of caring* composed of people who understand that part of the purpose of government is to act on behalf of the public interest and to provide those goods and services that people need but that private markets cannot deliver on their own unless they first distort and commoditize them. By linking these interests across sectors, it is possible to conceive of a broad social movement in support of democracy and public welfare that is powerful enough to build the institutions capable of countering the effects of commoditization. Religious and spiritual organizations and values may play a particularly important role here by insisting that nonmaterial and noncommodity values are ultimate and should take precedence over material values. The potential impact of spiritually motivated social action can be seen in the movement towards "voluntary simplicity"[4] or what might be called the decommoditization of everyday life. Spiritual values have been the foundation for many of the most important social movements of our time.

8.6 Government reforms for decommoditization

How should the six essential purposes of government change in order to offset the effects of commoditization? We will consider each one separately, although the functions clearly overlap. Commoditization's power as a selection process operates because of the natural tendency for investment of capital and other resources to flow toward that which can be most readily commercialized. Over time the imbalance favoring commodities grows. Government's power to build, regulate, tax, and spend can provide powerful counterbalance by shifting investment and other resources toward the underdeveloped noncommoditizable sectors of the economy, the economy of caring, the capacity for local self-reliance, the protection of ecosystem integrity, and the provision of the full range of noncommodity goods and services that cannot be expected to be provided by private industry.

Such a coordinated decommoditization strategy requires first and foremost a revisioning and revitalization of governance as the manifestation and agency of the collective interest in promoting the common good. An understanding of commoditization provides a common conceptual framework with which to design and implement public policies. The intent of decommoditization policies should always be to accomplish what private industry cannot accomplish and to regulate industries and markets to minimize the effects of commoditization on the environment and society. Thus, infrastructure development should focus on building up what private investments tend to ignore and commoditization tends to undermine. Once this becomes clear, the role of government grows more apparent, particularly in supporting low commodity potential research and development and stimulating investments in the economy of caring and relationship, and in dramatic improvements in the efficiency with which we use energy and materials to meet basic needs and build prosperous communities. These should be the three pillars of governance and public service: knowledge, care, and thrift. These should be the themes of all our reforms. First, some thoughts about how the government role in human rights and public safety would be affected by a decommoditization strategy.

8.6.1 Protecting human rights

The concept of what constitutes a human right would include the basic economic rights referred to in the International Covenant on Economic, Social and Cultural Rights.[5] It would also include rights to a healthy environment and the rights of local communities and indigenous peoples to control local commons, biological resources, and cultural destiny.

8.6.2   Defense and public safety

More effort would go into the low commodity potential services of community and neighborhood watch programs, family support, crime and violence prevention programs, conflict resolution, and others.

Although government's role in providing for public safety and protecting human rights is extremely important, the remainder of this chapter will focus mostly on the infrastructure, regulation, and taxing and spending functions of government.

8.6.3   Government infrastructure for the provision of public goods

Investments in infrastructure would be heavily weighted toward energy and material conservation, local community and bioregional support structures, environmental protection, and the institutions of caring and connection.

A key component of any sustainable development strategy is industrial policy designed to move industry in the direction of dematerialization, or the use and consumption of significantly less material and energy per unit of production. Many have argued for such a new industrial policy,[6] some calling it a new industrial revolution.[7] Although the importance of inventing new ways to get more out of less materials and energy cannot be overemphasized, if it is done without attacking the structural problem of commoditization it will not succeed. Unless the goal is shifted from improved efficiency (the most product with the least input) to improved sufficiency[8] (the most satisfaction from the least product), then dematerialization can never become decommoditization. Nevertheless, a new industrial policy is a critical aspect of the needed economic reforms. The components of such an industrial policy would be as follows:

- A major public investment in research and development of clean technology and the substitution of safe, abundant, renewable resources for hazardous and nonrenewable resources in all areas, especially in the energy industry.
- Major tax incentives and subsidies toward the development of zero emissions and integrated closed loop processes.
- Requirements that all products be designed for easy disassembly and reuse, including buildings.
- Strict energy efficiency requirements for all consumer end-products.
- Adjustments to current Best Available Technology rules to specifically promote the use of zero emissions technologies.
- Major new investments in rail service and other new low-impact transportation.
- Conversion of durable goods into services by shifting from retailing to leasing services for most major and even minor consumer goods and appliances.[9]

- The development of the physical services industries, particularly repair, restoration, cleaning and maintenance services, redesign and recycling services, and a range of other labor-intensive, hands-on services in conjunction with the new leasing industry.
- Major shift of public investment toward human services and social welfare.
- Establishment and maintenance of major bioregional research and development centers focused on LCP innovations in all areas, especially in appropriate technology, low-impact agriculture, energy and materials conservation, land use, urban community design and development, pollution prevention, and including education and extension programs to help individuals and communities gain greater levels of self-sufficiency in meeting basic needs.
- Expansion of the institutions of locally owned public utilities, employee stock ownership programs, and other means of democratizing capital.[10]

8.6.4   Provision of public goods and services

The only way to balance commoditization is for all those goods and services that are difficult or impossible to market profitably but are essential for maintaining quality of life in our communities to be provided by or subsidized by the community in some way. A successful decommoditization and sustainable development strategy depends on being able to shift time, attention, and money to the valuable goods and services that have low commodity potential. This will most likely occur in the public sector.

Most modern economies grow in large part through labor productivity improvements. But economies will only benefit from these labor-saving innovations in the long run if there are other opportunities to reemploy the newly redundant labor. In a commoditized economy, labor shifts to new or expanded jobs in the production of new commodities, which in turn expands the amount of energy and materials flowing in the economy; in other words, leads to economic growth. If, however, a growing portion of workers displaced by labor productivity improvements is shifted through public investment into the low commodity sectors of the economy of caring and relationship, then the freed labor will be valuably employed for the long term without necessarily increasing the flow of energy and materials.

Even without a deliberate effort to shift workers to public employment, in every growing economy the public sector expands as a proportion of GNP, even under the pressures of commoditization and without any real increase in the amount of attention, time, and resources allocated to the noncommodity public services. It is important to understand why the public sector paradoxically grows under commoditization. This has to do with the pressure on wages of increases in private sector labor productivity. When industries lower costs through labor-saving innovations they become more profitable. The workers that remain, although reduced in numbers, can command higher wages. In general, public sector workers, through their labor organizations, attempt to keep pace with private wage scales. However, public employees are generally in the lower-commodity services sectors (or they would not be *public* employees) and cannot increase their productivity to the same degree as is possible in the private sector. As a result of this natural process, there is an increase in the proportion of wealth allocated to the public sector. In the same way, most public sector employees do not produce goods and services that are counted as part of GNP, and therefore the public sector expenditures as a portion of GNP also increases.

A decommoditization strategy would increase public sector expenditures as a proportion of GNP considerably more. Such an outcome can only happen through the collective expression of nonmarket values through politics and government, not through markets. Markets can and must be enlisted to allocate resources efficiently among alternative commodities, but the survival of nonmarket goods and services in the economy of caring and relationship depends on the capacity of public institutions to carve out economic niches where these goods and services can flourish. As politically difficult as it may presently appear to be, sustainability and the health of the planet depends on government — big government for big problems, small government for small ones. The issue of democracy, fairness, and public accountability become, therefore, the most urgent of the sustainable development issues.

If strong, effective government is necessary, then it must also be competent, fair, and responsive. Thus, the movement for environmental sustainability must be a movement for good government, strong government, and the public capacity to balance commoditization with the collective provision of important public goods and services. In this need lies a potentially powerful alliance between advocates for environmental protection, good government, and public service unions and professional associations. The potential social movements to reclaim governance and the capacity to act on behalf of the environment are discussed at the end of this chapter.

8.6.5   *Government regulations for environmental protection, commerce, trade, and land use*

The modern regulatory apparatus in industrialized nations is usually justified on the basis that it promotes the common good in cases where free markets fail to do so on their own. For instance, health insurance companies would prefer to insure only people at low risk of illness, but the state regulates insurers to some extent to make certain that the risk and cost of health insurance is spread more evenly throughout society. The common good comes at the expense of some percentage of profits that insurers must forgo. Insurers benefit in the end by the increased stability accorded to the insurance market from the reduction of the numbers of uninsured. In theory, government regulation should only be objectionable when it interferes with markets that would more effectively serve the common good if left to themselves.

The question then is not whether to regulate certain industries but how to regulate them. Commoditization operates to favor certain kinds of regulation, those that resist commoditization the least. First, regulations are deliberately designed to have the least negative impact on economic growth and in the case of commerce regulations are designed specifically to encourage economic growth. As we have seen, encouraging economic growth in a commoditized economy necessarily intensifies the force of commoditization. Furthermore, government, as much as consumers, is subject to the preference for commodities when selecting solutions to any particular need. It is simpler to buy something than to do something, especially when you are surrounded by people eager to sell.

It is also simpler to regulate in a piecemeal fashion rather than to have a coordinated strategy. This is as true for environmental regulation as any other kind. Air quality statutes are passed in response to air pollution, water quality statutes in response to water pollution, and so forth. There is no overall, aggregated measure of environmental quality, and there is no holistic policy of reducing effects of human activity on the biosphere. The establishment of the regulatory regime in most industrialized nations in the 1970s took this approach and subsequently created a huge pollution control industry. Industries took the easiest, most commoditized solution by purchasing add-on commodities in the form of pollution control devices that brought them into compliance with discharge regulations. Pollution control technologies do indeed reduce the emissions of regulated pollutants into air and water, but environmental quality may still suffer long-term damage because filters and other emission control technologies merely move pollutants from one medium to another (e.g., from the air to the soil when filters are disposed of), and because the production, maintenance, and disposal of pollution control devices often lead to *increases* in total energy and material use. Meanwhile the devices allow polluting industries to continue to expand while reducing the worst of the localized and visible pollution in the wealthy industrialized world. Pollution has also been exported, as the dirtiest industries are relocated to the poor, newly industrializing parts of the world.

One alternative to piecemeal pollution control would be a coordinated strategy of waste reduction and pollution prevention for each major economic sector aimed at decreasing the total amount of energy, raw materials, and pollution produced in each sector and the economy as a whole. A regulatory regime to accomplish this would necessarily be quite demanding, requiring polluters to redesign production, distribution, and waste disposal systems while at the same time using the government's powers to tax and spend to discourage consumption of products with high material and energy costs.

Environmental regulations will be much more effective when accompanied by regulatory reforms in other areas that encourage the decommoditization of the economy. Most estimates conclude that in order to reduce the human impact on the environment while continuing to grow economically, we must become four to ten times more productive in our use of physical materials and fuels.[11] Although this may be a difficult technological challenge, it is instructive to note that modern industrial technology and commoditization have increased the productivity of labor by more than 100-fold during the period of industrialization. Each hour of human labor can produce more than 100 times more economic goods and services than the same hour could 100 years ago.[12] Decommoditization policies that shift the incentives for investments toward materials and energy productivity can stimulate similar productivity gains for these inputs.

Improvements in energy and materials efficiency alone will not, however, lead to general decommoditization. Every savings in energy and fuels represents cash not spent. If that cash is simply shifted to purchasing other commodities, then no matter how energy- and materials-efficient our economy becomes, overall use and consumption of energy and materials is likely to continue to rise. This is clearly demonstrated in the economic modeling done by Faye Duchin, who demonstrated that, given current projections in economic and population growth, even assuming a scenario in which both the industrialized world and the less industrialized poor nations adopt the most clean and efficient technologies available today and conserve energy and recycle raw materials at the levels achieved by only a few countries at present, "despite substantial savings in energy and materials that are attributable to the new technologies... the pollutants that are tracked in the model all continue to grow, rather than fall, over the period studied."[13]

Policies directed to promoting efficiency improvements must be accompanied, therefore, by decommoditization policies that actually slow economic growth while increasing public welfare. This can only occur if the money generated by reduced costs goes in two directions: (1) toward increasing capital accumulation through an increased savings rate, and (2) the investment of that capital toward improving public welfare through investments in supporting local self-reliance, building the infrastructure of caring and connection, and protecting ecological integrity. Success will require reforms in the rules governing corporate practice, public and private credit and investment, and the allocation of government revenues. Obviously this cannot be accomplished by current environmental regulatory agencies alone. The only way for a green economy to emerge is for ecological principles and the reforms they imply to permeate all levels of government and policy, including governance institutions at the global and ecosystem levels which do not presently exist.

Land use decisions have enormous environmental consequences. The problems associated with attempts to steer land-use toward more sustainable, environmentally friendly patterns raises many of the same sorts of issues of government regulatory capacity. Most land use regulations, zoning, and land development policies have some decommoditizing effect on property — one's property rights are necessarily constrained by them to some extent. This is particularly true in cases where communities attempt to regulate and control suburban growth or constrain and direct urban growth to patterns meant to best serve goals such as transportation efficiency or clustering of public works services. Some communities have established urban growth boundaries beyond which housing and commercial development are not allowed. The stricter the land use policies, the more they come into conflict with commoditization. This offers another potential link in the building of political alliances against the harmful effects of commoditization, particularly between advocates for the revitalization of inner-city neighborhoods, promoters of public transportation, and protectors of green space and wild areas from the effects of suburban sprawl.

8.6.5.1   *Trade*

Trade policy is invariably put to the service of the globalization of commoditization, and the force of commoditization in turn stimulates more and more international trade in a global marketplace. Commoditization is exactly the preference for things that can function well in a free-trade environment. Free trade and commoditization feed each other. The argument for free trade rests on the logic of comparative advantage: if a country exports those goods it can produce most efficiently because of its particular endowments of natural resources and the skills of its people and then imports the things it needs that it is less able to produce efficiently, then the net result will be increased efficiency and therefore increased wealth for everyone. The advantage to local consumers is the greatly increased availability of goods and services for purchase and decreased prices resulting from increased competition. In the current trade climate, all nations are encouraged and, in many cases, compelled to open their domestic markets to imports and to produce more for export to earn foreign currencies so they can participate even more in the global markets.

The logic is impeccable, but it does run into some predictable problems. For example, as developing countries around the world all increase exports, particularly of primary materials, prices drop, undercutting many of the benefits derived originally from their comparative advantage in primary commodities. Those who gain from this situation are the wealthy consumers who benefit from falling prices and the investors and manufacturers of consumer goods who gain the windfall both from falling prices of raw materials and the freed up consumer cash in the rich countries that results from lower consumer prices.

The problem with the logic of comparative advantage is that it accepts as a given both the accidental inequalities of geography and the deliberate injustices of history. It also assumes that all nations interact with a free market from an equally competitive bargaining position, eager to exchange their most efficiently produced goods with the rest of the world. Many nations' and peoples' competitive advantage derives from their position of weakness. They are poor and able to work for lower wages; therefore, low labor costs are their comparative advantage. They have raw materials but not the capacity to process them into useful goods. Thus, the export of primary materials is their comparative advantage. But the status of workers and the lack of local industry is as much or more a result of colonialism and the history of oppression as it is that of geography and cultural development. This means that people are asked to compete in a situation where the misery of their own oppression is exactly what they have to sell.[14]

The logic of comparative advantage also leads invariably to specializing in those export commodities for which a nation has some perceived advantage. This specialization, as it intensifies, necessarily reduces local economic diversity. With loss of diversity comes increased vulnerability to competition, fewer options, and lower resilience in times of economic stress. As in every other area we have discussed in this book, the challenge is to maintain the balance between the pressures of commoditization and other noncommodity values.

The globalization of commoditization and accompanying concentration of capital in the hands of transnational corporations leads to increasing marginalization of the majority of the world's people who have little or nothing to contribute to the cash economy. As commoditization shapes the global economy and absorbs resources, more and more of available human time, attention, and natural resources are shifted toward satisfying the consumer demands of those who can pay. The gap between the rich and poor will grow ever wider unless the forces of commoditization are counteracted. The United Nations Human Development Program reports that in 1998 the richest 225 individuals had as much wealth as the total income of the world's poorest two and half billion people.[15]

Gaining some social control over the flow of international trade and investment is necessary and will require new global institutions with substantial authority to regulate international commerce to protect the public interest. Nations are increasingly unable to control or even influence the direction and content of global commerce.[16] Small communities and neighborhoods are even more disadvantaged. Seeing that renewed investment in local communities and appropriate technology will be necessary to make the transition to a sustainable economy, and that these tend to have low commodity potential, the conflict between the interests of sustainability and free trade will only grow stronger. At some point in the expansion of the global economy the need will emerge to counterbalance the forces of commoditization at the global level. As the institutions of the global economy are built, global institutions must also emerge to protect the common good. There is no way around the counterintuitive fact that the empowerment of local communities and the decommoditization of the economy necessarily require new and powerful authority placed in global trade institutions that are democratically accountable to the peoples of the world.

At present all the trends of trade and investment are in exactly the opposite direction. There is much less public than private investment in the so-called emerging markets. The amount of private capital flowing into these markets rose from $44 billion at the beginning of the decade to an all-time high of $244 billion in 1996 (in current dollars). At the same time, publicly funded spending, or "official development assistance," fell by more than 25%. In 1990 less than half of the international capital moving into the Third

World came from private sources; by 1996 this share had risen to 86%. These types of funds, particularly under the conditions of commoditized capital flows (meaning mobile and speculative), are highly vulnerable to mood swings and sudden changes of fortune. This was demonstrated in 1997, with the onset of the global economic crisis in Asia, when the trend reversed and private capital flows to emerging markets fell by around 20%.[17]

The volatility of capital flows in the late 1990s initiated a period of severe economic uncertainty for most developing countries. This economic uncertainty strictly limits the capacity of government to protect natural resources and the environment. Globalization of economic development and investment makes it nearly impossible for weak governments and underdeveloped economies to act alone to protect their economies from capital flight and currency speculation. The economic strain places tremendous stress on national governments to initiate or continue rapid, unsustainable development by accelerating the exploitation of natural resources as the only hope of offering investors the potential for quick fortunes, and thereby attracting foreign capital investment. The resulting impact on the environment has been devastating. Since no individual government is now capable of effectively controlling capital movements, the only hope is through internationally negotiated capital controls associated with international trade agreements.

Investment in less developed and less commoditized countries takes three forms: (1) foreign direct investment, in which foreign capital is invested in the productive capacity of a country — often this is done in joint partnerships between the foreign investor and a domestic partner; (2) portfolio investments, in which foreign investors buy stocks and bonds issued by domestic markets; and (3) loans from foreign commercial banks, either to local or foreign investors for use in the domestic economy. Each of these has grown much more mobile and volatile at the same time that the availability of official, government-to-government economic assistance has declined. The result is the dwarfing of publicly financed international development assistance by private investment and the resulting intensification of commoditization. It might be possible to counterbalance this by

- Establishing international controls over capital mobility through international agreements.
- Levying an international tax on capital movements and currency exchange to provide resources for the creation of international institutions for market regulation, environmental protection, and labor and other social standards.[18]
- Requiring that all trade agreements be accompanied by civil charters covering environmental protection, workers' rights and safety, food safety, human rights, and other common goods neglected because of commoditization.
- Supporting local and regional self-reliance and reduced dependence on the global economy.

- Reforming trade agreements to control capital movements, and applying tariffs and duties which encourage optimum local self-sufficiency in food and energy and the protection of natural resources and the environment.

8.6.5.2 Environmental protection

The body of laws and regulations that presently govern the activities of business, industry, local communities, nonprofit organizations, and individuals has over time been distorted in the direction of commoditization. New rules and regulations are needed to counteract this distortion and to create the conditions under which the natural world is protected without sacrificing prosperity. This is the heart of the challenge of sustainable development. In general these regulations would increase the role of government in economic regulation and would increase employment in conservation and preservation. They would force producers to design products that are energy and materials efficient. Such initiatives and reforms might include a shift in the principle underlying environmental protection regulations from pollution control and reduction to pollution prevention and elimination. This will require

- Gradually declining allocation of total maximum wasteloads in a given watershed or airshed, with concomitant public capacity to monitor releases and require necessary reductions.
- Requirement that all major manufacturers develop and implement both site-based and industry-wide energy and material reduction plans.
- Redesign of production processes so that all waste, including waste energy, is captured and used.
- Requirements for manufacturers to take back their products for reuse or recycling, in effect requiring expansion of leasing arrangements for most major purchases.
- Institutions and incentives for the exchanges of discharge rights and waste materials operating under rationing systems, with strict caps on overall pollution.
- A permitting system for all new synthetic chemicals and materials with permits allowed only after clear demonstration of absence of environmental harm.
- Government capacity to regulate by classes of compounds or types of processes that are reasonably anticipated to have negative environmental consequences.
- Regulatory structure guided by sustainable development principles, using cost/benefit analyses that include the nonmarket value of social and environmental goods and services.
- Meaningful environmental assessments incorporated into the regular functioning of all government agencies including the permitting functions associated with all major private investments. These will be based on broadly defined costs and benefits, including nonmarket ones.

8.6.5.3   *Corporate influence*

Corporate power and influence naturally grow along with commoditization in a classic positive feedback effect. Corporate power is used to promote policies that spur commoditized economic growth. As these policies cause the economy to become more and more distorted toward commoditization, corporate power grows even more. The more corporate power grows the more corporations influence public policies at all levels toward the encouragement of commoditization, and so on.

In the most recent historic period this has been most noticeable in economic globalization. Major transnational corporations have used their influence to convince nations to lower barriers to investments, reduce or eliminate regulations on taking profits out of countries, and have encouraged nations and regions to compete with each other to lower corporate tax rates and increase publicly supported incentive packages designed to attract foreign investment. Speculative investors have achieved the elimination of fixed exchange rates and removed most barriers to the movement of currencies across borders. The whole process has been facilitated by the increasing commoditization of finance and information through the computerization of information and financial transfers among banks, stock markets, and commodity markets. While the infrastructure of the global economy has matured and come under effective control of transnational corporations, the institutions of international civil society and governance capable of regulating corporate behavior for the common good have matured slowly, if at all.

Amid the antigovernment rhetoric of much modern political ideology, it's easy to forget that the private corporation is an artifact of governance. The history of corporate law is readily understood in terms of the logic of commoditization. The whole point of corporate law is to create conditions conducive to amassing capital and establishing commercial organizations. The existence of corporations and the body of corporate law is a public good created by collective action. Therefore, collective action, through the exercise of governance, can and should be applied to reigning in the power of corporations. No strategy for sustainable development or correction of the economic distortions caused by commoditization can possibly succeed without doing so. The following policy directions would begin to check the continuing growth of corporate power:

- Reform of all government agencies to prevent capture by industry. This would include the following:
  - Hiring preferences given to individuals from outside the regulated industries and particularly to those with backgrounds in the public or nonprofit sector (greatly expanded as a result of this overall reform package).
  - Establishment of sustainable development policy goals in the areas of increased materials and energy efficiency in the context of rationed

total use, increased employment, and reduced waste for every agency including military, transportation, commerce, housing, etc.
   – Preferential treatment for small and local business in all government services.
 • Reform of corporate law to weaken the power of corporations by
   reviving the notion of corporate charters as privileges issued on behalf
   of the people by their representatives.
   – Requiring periodic review of charters, with the possibility of revo-
   cation for irresponsible corporate behavior.
   – Limiting the legal fiction of personhood which grants corporations
   the same civil rights as individuals while at the same time granting
   limited legal liabilities that would never be given to individuals.

8.6.5.4   Economic development and modernization

The history of economic development and modernization has been deeply
distorted by commoditization. Those aspects of development and modernization that served commoditization received abundant investments of time,
human attention, and resources, and those aspects meant to build the capacity for delivering public services and support the noncommoditizable req-
uisites of a good life grew increasingly underdeveloped in comparison. The
problems associated with this distortion of development are easy to observe
even if difficult to control. These problems grow more or less severe depending on the state of economic development in any given time or place.

   The problems lead to periodic reevaluations of economic development
strategy and calls for reform. The most recent reform movement has gathered
around the critique that describes conventional economic development practices as *unsustainable* because its unintended consequences damage the environment and weaken local community stability and resilience. In response a movement for *sustainable* development has emerged. As our analysis shows, the
recent crisis of development is an outcome of the process of commoditization
and cannot be mended without a deliberate strategy of balanced development.

   This type of development has sometimes been linked to "mid-range" technologies in the development literature to distinguish them from both the high-
cost, large-scale, capital-intense mega-projects and low-tech, labor-intense subsistence tools available in underdeveloped economies. What are called mid-
range technologies are consistent with what I have described as technologies
with mid-range commodity potential. They are also consistent with the types
of technology that have emerged from the appropriate technology movement.

   Many international development agencies and major philanthropic foundations have recognized the need for balanced social, economic, and political
development. Over the years many critiques of international development
policies have been leveled, and the development community has entertained
many versions of the call for reform, most recently in the guise of sustainable
development. In the present era, with the private flows of capital and investment now far outweighing that provided by international development agencies, the need to counterbalance the forces of commoditization is greater than

ever. The currently most powerful institutions for global economic development — the World Bank, International Monetary Fund, and the World Trade Organization — are all committed to a strategy of increased commoditization. The policy formulas and so-called structural adjustments that developing countries are being pressured to adopt are for the most part the exact opposite of a decommoditization strategy. Governments are pressured to open their doors to free movement of capital, to lower domestic taxes, and greatly reduce public spending on welfare and social services. The objective is to shift money and resources to the most commoditized sectors of the economy in order to stimulate economic growth in terms of commercial goods and services. This is done partly to raise the funds needed to pay debts to the world's major lenders, further concentrating wealth and power in the hands of the wealthy and further stimulating commoditization in the classic positive feedback process.

The entire institutional structure of global economic development must be reformed in order to gain some control over this runaway commoditization feedback mechanism. An international sustainable development infrastructure is required beginning with a commitment from the major international development agencies to support a network of well-funded bioregional research and economic development centers specifically devoted to local sustainability, the development of diverse, self-reliant economies based on local ecosystem characteristics, local renewable resources, local crafts industries, urban–rural economic linkages such as community-supported agriculture and urban gardening, renewable energy and public transportation. Such an infrastructure could be funded by taxes and fees on currency exchanges and other international trade as well as on global taxes on carbon emissions and the trade of carbon emission rights.

8.6.6 *Taxation, public spending, and the supply of money*

Government's ability to raise revenues and to spend those revenues gives government the capacity to shape the economy. Because taxes can shift the allocation of resources in an economy, a government's authority to tax and spend may be the most important power in terms of reinforcing or counteracting commoditization. Taxes and public spending also have an effect on the availability and cost of private credit. Credit plays a major role in commoditization, especially in its acceleration in advanced industrialized societies. These issues are discussed below.

8.6.6.1 *Tax policies*

Tax policy should be used to counteract the effects of commoditization.[19] The aim of a strategy to correct the distortions of commoditization would be to decrease the effective price of labor and increase the price of raw materials and energy by gradually shifting taxes away from wages and income and onto materials and energy. Tax policy may be the most important and efficient tool for counterbalancing commoditization, because it directly influences prices and makes it possible to shift money from the most commoditized sectors to the least. Such taxes could create incentives for producers to reduce the material and energy content of their products, thereby reducing the environmental impacts of production. Taxes could also be targeted to the key elements of commoditization such as packaging, advertising, international trade, capital gains, and currency exchange. The revenues collected from such taxes could be used for targeted public investments in further production efficiencies and community-building.

Raising the price of energy by imposing energy taxes would probably be the most efficient and effective means of decommoditization. It could accomplish much environmental good by itself and be relatively simple to administer if not to legislate. Energy taxes could be made more sophisticated and be targeted for particular environmental purposes, such as a tax on carbon content to reduce the emissions of CO_2. Such taxes would encourage the shift from fuels with the most carbon, coal, toward natural gas and ultimately to carbon-free hydrogen fuel.

Shifting the tax burden from labor (in the form of income and payroll taxes) to energy and materials (in the form of raw material and fuel taxes) would lower the effective cost of labor in comparison to energy.[20] Human labor is the least commoditizable of the factors of production, and commoditization always moves in the direction of increasing labor productivity through substitution of energy and energy-intensive capital for labor. Raising the price of energy through fuel taxes and simultaneously reducing the price of labor through lowered income and payroll taxes shifts the potential gains from productivity improvements from labor to energy and materials, thereby moderating one of the most potent stimulants for commoditization. With labor costs declining in relation to energy and raw materials, it begins to make sense to repair rather than replace consumer goods. Retrofitting and insulating homes becomes less expensive than paying for fuel for heat. Labor-intensive organic farming practices begin to compete with energy-intensive, high-input agriculture. Locally hand-made crafts begin to compete with imported Barbie dolls.

In addition to taxing materials and energy, it may be possible to deliberately tax pollution. Taxes could be levied on air, water, and soil pollution of all kinds, with the most toxic and damaging pollutants being taxed the most. Taxing pollution output is another way of internalizing costs of production that are usually externalized to society. Such environmental taxes are already in use in countries around the world, and their popularity is expected to rise as the costs of pollution are better understood.[21]

In the end it makes sense (as it has for such "sin" taxes as the levy on alcohol, tobacco, and gambling) to tax things we would like to reduce so that the effect of higher prices can have social benefits. When taxes are used in this way, they are sometimes referred to as "regulatory" taxes. Regulatory taxes work best when levied against activities that are elastic; in other words, people are able and willing to change behavior in order to avoid the tax. It makes sense therefore to accompany regulatory taxes with incentive programs and infrastructure development to encourage the desired behavior. Energy and materials taxes will work best in conjunction with an industrial policy that encourages efficiency and a public investment and social welfare program that promotes sustainable development education and alternatives. This will make it possible for people and industries to avoid the tax by shifting the types of goods produced and production processes used. Of course, a government that relies too heavily on regulatory or "sin" taxes runs the risk of being in a serious conflict with itself over reducing those activities and thereby reducing government income. Regulatory taxation must always be balanced with strictly revenue-generating taxes such as sales taxes and taxes on various forms of income and property.

Consumption taxes are an alternative or corollary approach to energy and raw materials taxes. Of these, the value-added tax may be the most powerful. A value-added tax would impose a tax at each stage in a product's manufacturing life at which value is added to it. This would have the effect of raising the cost of the product and thus inhibiting consumption. Unlike the sales tax, which simply adds a fixed percentage to the cost of buying products, the value-added tax imposes greater costs on highly commoditized goods. In other words, a product that is highly processed — i.e., that undergoes many stages of added value — is taxed more than a product that is only minimally processed. So instead of just discouraging consumption by raising prices, value-added taxation could encourage some substitution away from highly commoditized goods and services.

One objection to the value-added tax is that it unfairly burdens the poor, who must consume most or all of their income, as against the rich, who consume only a small part of their income and would consequently receive a huge tax cut from the reduction of income taxes with no concurrent imposition of another tax. In other words, using value-added taxation instead of income taxation would shift the tax burden from the rich to the middle and lower classes. Consequently, some income tax should remain in place, especially on investments. Moreover, it would be necessary to provide a tax credit or refund to the poorest citizens so that they do not pay a disproportionate share of taxes. Finding an equitable mix of income, value-added, and pollution taxes as well as tax credits and refunds for the poor is an achievable task.

Other proposed tax polices include the following:

- Accelerated depreciation allowances for old energy-inefficient plants and equipment to encourage energy-efficient replacements.
- Employment tax credits and elimination of local "head taxes" (per employee taxes often levied on businesses by local governments).
- Tax incentives for emissions reductions, recycling, and other environmentally beneficial private initiatives.
- Tax credits for worker training in the new skills needed by an energy efficient economy.

8.6.6.2 Expenditure policies — subsidies and government purchases

Tax policies, although necessary, are not sufficient to encourage a balanced economy. How tax revenues are spent has an even greater potential to either increase the distortions of commoditization or help to balance them.

If goods and services with high value and low commodity potential are going to have any chance of economic survival, then some methods have to be devised to transform their high social value into actual support, financial and otherwise. This is the very essence of subsidies, where government provides resources to correct the negative side-effects and distortions created by free markets. If a government acts primarily to promote the policies of economic growth, then government purchases and subsidies will further serve the cause of commoditization rather than correcting the distortions of commoditization. This problem becomes extreme when powerful economic interests so capture government that its policies no longer serve to balance the distortions and correct the externalities of free markets, but instead aggravate the distortions by subsidizing the already privileged. In his book, *Paying the Piper: Subsides, Politics and the Environment*, David Roodman describes the economic distortions and environmental damage created by such subsidies.

Governments around the world spend upward of $500 billion on subsidies for agriculture, forestry, mining, and many other industries, affecting the economics of these industries and their environmental costs. These types of subsidies make commodities out of natural resources that would otherwise be too costly to commoditize through the workings of the free market. They intensify rather than correct the effects of commoditization. The U.S. government currently favors the fossil fuel industry with about $36 billion per year of subsidies, which have the effect of keeping fuel prices low and encouraging wasteful energy use by Americans.[22] Estimates vary, but if fuel prices in the U.S. were based on the real cost of production, a gallon would cost about $2.50 ($0.70 per liter). If the cost of air pollution, highway construction, traffic management, and other publicly funded transportation-related expenses were also added in, the real cost would come out at around U.S. $4.50/gallon.[23] Even without the imposition of the energy taxes discussed above, these prices would begin to shift the economic calculations of investment decision toward energy efficiency.

Even when subsidies are instituted for ostensibly good reasons, their unintended consequences can be very damaging. For instance, agricultural subsidies are intended to protect farmers from market instability, ensure an agricultural base to protect a nation against dependence on food imports, and to provide a predictable food supply with stable prices. In practice, however, the largest share of agricultural subsidies go to the minority of large farms that produce the majority of the food.[24] The losers in this arrangement have been small to midsized farmers, who cannot compete with their oversubsidized, oversized neighbors. The subsidies encourage ever-expanding monocultural production, because companies and farmers know the government will buy their products. As a result, more acres are put under the plough, with the concomitant destruction of habitat and pollution of soil and water from agricultural chemicals.

What needs to be subsidized are exactly those goods and services we have discussed throughout this book, the systems of health promotion and illness prevention, of ecological health and integrity, of local community self-reliance, of the economy of care and connection. Because they resist commoditization, they cannot survive without subsidies derived from the collective resources of people, either through government or through voluntary associations and civil society groups.

There are several ways for governments to provide resources for desired social and environmental purposes and each has a different commoditization potential:

- The direct provision of these goods and services, where appropriate, has the most potential for decommoditization by reallocating society's resources directly from the tax-paying commercial sector to public goods and services for which markets rarely exist.
- Government can also choose to allocate revenues to the private sector, but with the intention of supporting particular kinds of business, such as small organic farms, neighborhood service providers, housing co-ops, etc. In the U.S. the Small Business Administration was created to serve a particular clientele in this way. By supporting small private providers of LCP goods and services, government limits the amount of direct service it provides while still serving the purpose of decommoditization and increased employment.
- Government can choose to allocate its resources to support large and powerful corporations in the hopes of stimulating economic growth and employment. This is the tack taken by most industrialized nations in the present period; the one that produces the fewest jobs per capita outlay and the one that accelerates commoditization the most.

It may be possible to institute a "green" economic or industrial policy without explicitly addressing commoditization. Most of the current green taxes and subsidies are focused in industrializing appropriate technology. Denmark is one of the leaders in this area. Fuel taxes have been used to subsidize a growing wind power industry to the point where the Danes now produce 80% of the world's wind turbines and export 90% of those they build.[25] This model, the development and production of renewable energy and other clean technology equipment and supplies and the subsequent export of these goods to a world set on a clean energy path, is the commoditized version of sustainable development. It is necessary if the transition to a more energy- and materials-efficient economy is going to occur, but it is not sufficient to alter the economic system dynamics driven by commoditization that underlies the environmental crisis to begin with. All attempts to create a green economy or a green industrial revolution, although extremely important, will find themselves faltering without an accompanying decommoditization strategy intended to modify the system dynamics.

8.7 Credit policies

Credit is, in effect, borrowing from the future. This makes a lot of sense in the case of large purchases, such as a home, when the benefits are spread out over a long time into the future, in the case of building infrastructure that lasts, and in the case of capital investments that are expected to produce income in the future needed to pay off the debt. Credit does not make sense as a means to support current consumption except under special conditions of necessity. Under the strains of commoditization, when almost all human satisfactions are gained through purchase, borrowing for present consumption grows increasingly popular and the economy becomes dependent on the stimulus provided by easy credit. But debt obligations limit the choices individuals and societies have in making decisions on how to allocate their resources in the future. This constrains future opportunities and locks society into a pattern of never-ending economic growth. But as we have seen, as long as economic growth means the increasing mobilization of energy and materials, then ecological constraints will eventually make growth undesirable if not impossible and future debts unpayable. This is a formula for economic collapse and all its attending misery.

The difficult transition to an environmentally sustainable economy will require changing the patterns of production and consumption and, prior to disaster, shifting investments toward those goods and services that provide the greatest satisfaction for the least expenditure of energy and materials. But as we have seen, these tend to be the locally produced goods, culturally and ecologically appropriate tools, and the services of mutual aid and community building, all of which have some but not large amounts of commodity potential. The transition to a sustainable economy, therefore, will require considerable new investments in things that will not produce large profits on investment. Economic growth cannot be the goal of economic policy. Economic policy will need to be directed toward goals and objectives determined by the vision of a sustainable society. This will require unprecedented levels of intervention in economic decision-making, starting with the allocation of credit toward specific, socially beneficial ends at very low or no interest. Society will need to gain more control over the availability and price of credit. As we move toward a society built on smaller units of local communities and neighborhoods of mutual aid and relationships of caring with each other and the Earth, we will need large numbers of local, small creditors accountable to the local community and neighborhood. This is, of course, the vision of the early credit union movement.[26]

Government policy in most modern societies affects the availability and price (interest rates) of credit through its taxing, spending policies, and its control over the money supply. These and other tools will need to be directed toward the goals of a sustainable economy if we are to have any chance of success. The difficulties will be enormous and the challenge great.

8.8   Building a movement

Statistics about how we presently spend our resources are often shocking.
The World Watch Institute, a Washington, D.C., sustainable development research organization that tracks environmental trends, publishes
a section in its *World Watch* magazine called "Matters of Scale." A recent
issue (January/February 1999) gave readers the series of juxtaposed figures shown in Table 8.1, drawn from the United Nations *Human Development Report*.[27] These and similar statistics are often cited to demonstrate
the greedy and selfish nature of human beings and the perversity of our
priorities, but a closer look shows something else quite clearly. Without
deliberate intervention, the flow of resources concentrates in the economy
of commodities and abandons the economy of caring and connection. Per-
fumes are a perfect and simple article in the economy of commodities, the
provision of health care is a complicated system in the economy of caring.
The same contrast is true for sanitation vs. cosmetics and pet foods vs. the
provision of health and nutrition. Military expenditures largely go to
weapons and the soldiers who use them. It is much simpler (and more
amenable to commoditization) to destroy things than to build them. Education is one of the least commoditizable goods in the economy of caring
and connection. Finally, as we have seen, commoditization leads to
increasing concentration of wealth and power in fewer and fewer hands
and the increasing marginalization of the poor. It is only possible to

Table 8.1 Comparative Spending

Amount of money spent annually on perfumes in Europe and the U.S.	$12 billion
Amount of money needed each year (in addition to current expenditures) to provide reproductive health care for all women in developing countries	$12 billion
Amount of money spent annually on cosmetics in the U.S.	$8 billion
Amount of money needed each year (in addition to current expenditures) to provide water and sanitation for all people in developing nations	$9 billion
Amount of money spent each year for pet food in the U.S. and Europe	$17 billion
Amount of money needed each year (in addition to current expenditures) to provide basic health and nutrition needs universally in the developing world	$13 billion
Amount of money spent each year on militaries worldwide	$780 billion
Amount of money needed each year (in addition to current expenditures) to provide basic education for all people in the developing world	$6 billion
Combined income of the world's poorest 2.5 billion people	$1 trillion
Combined wealth of the world's richest 225 people	$1 trillion

Source: Worldwatch Institute, *World Watch*, June/July 1998. With permission.

change this situation, and to reverse the assault on global ecological integrity, through a collective decision to provide a balance between the economy of commodities and the economy of caring and connection.

Outrage over the results of commoditization and the desire for change is a natural reaction to the blatant misallocation of resources in the world. Many people respond personally by changing and simplifying their own consumption patterns, voluntarily serving others, and spending their lives in the economy of caring and connection by working in education, alternative health care, organic farms, etc. As important as all these responses are, they will always be marginal without fundamental change in the structure of the economy that privileges commodities over common and public goods and services. The public sector must grow in relation to the private sector, and the economy must be treated as a subsystem of the world's natural and social systems.

If an economy of commodities is allowed to serve its own ends, then commoditization and its attendant consequences are inevitable. The economy must be made more responsive to the desire of human beings for a good life in sound, sustainable communities on an Earth that is ecologically vibrant and healthy. These changes will require a major new political movement to assert the values of caring and connection distinct from the values of commerce.

What our analysis uncovers is the deep connection between such seemingly disconnected issues as pollution, oppression of women, appropriate technology, the rights of indigenous people, working class oppression, grassroots democracy, public transportation, the low pay of service workers, the isolation of the disabled, globalization, and many others. Although several efforts have been made to bring people concerned about these disparate matters together into a progressive majority, a common political agenda and strategy has not been clear. With the decommoditization strategy outlined in this chapter, we begin to see our way out of the trap of commoditization and into new forms of governance and popular democracy. The opportunities for coalition building and real political and economic change are exciting.

They begin with the recognition that an equitable and sustainable society is a clear and desirable possibility for which it is worth working and organizing. There is by now considerable evidence that dramatic improvements in energy and materials efficiencies are practical and possible. There are large differences between advanced economies in terms of the amount of energy used per capita GNP: the average person in the U.S. uses 2.3 times as much fuel as the average European, and no one claims that the quality of life in the U.S. is more than twice as high as in Europe.

Progress in energy efficiency has been made largely through technological change and only relatively minor policy incentives. With real commitment to invest in the new industrial revolution, these gains can be accelerated. The commodity economy can readily be made more efficient and earth friendly. With the right policy signals and reforms that encourage conservation, the transition to efficient economy will take off.

The major oil companies, the energy, automobile, and a few other industries will resist the change, but resistance can be overcome through an alliance of environmentalists, consumers, and the majority of industrial groups. These groups will understand what Amory Lovins claims, that global warming and other environmental problems are not threats to business; rather, they are huge opportunities to benefit from a major new wave of investment in plants, equipment, appliances, and the other parts of the commodity economy that will be retrofit or replaced.[28] The efficiency revolution will mean new jobs. Every dollar spent on efficiency produces more jobs than a dollar spent on energy production. The first and simplest demand, therefore, is for a global political commitment, following the global warming treaty, to accelerate the existing trends toward energy conservation and efficiency. Public concern about global warming and its effects, along with the spread of information and examples of how to design and produce for super efficiency, will provide the catalyst for change.

The benefit of promoting this vision of a new industrial revolution or an efficiency revolution is that it is a hopeful picture of the future that can be juxtaposed against the sense of foreboding with which many people are looking into the future. It is doable and politically feasible given the natural conjunction of business, consumer, and environmental interests. It is the first step in a decommoditization strategy, because it can be popular and at the same time require increased government involvement in markets through public infrastructure investments, tax and subsidy policies, and regulations so that prices reflect the environmental and social costs of products. It will unleash a tremendous amount of creativity and enthusiasm from designers, scientists, and tinkerers of all kinds. The employment benefits will bring portions of the labor movement who will benefit and create a political demand for a just and fair transition for the workers in the old industries where jobs will be lost. This creates the need to assist workers and communities presently dependent on old, energy-inefficient industries to more smoothly make the transition to the new jobs and opportunities.

Reform of the commodity economy will be the first successful battle in the movement for sustainable development, and it will greatly enhance the possibilities of success in later struggles, but it will not succeed by itself, because it will not likely take on the job of providing ongoing economic counterbalances to the pressures of commoditization. There are many flaws in the argument for the new industrial revolution, not the least being that retrofitting of the present economy at the scale that will bring about 50% or 90% improvements in energy efficiency will require massive new construction and new production, all of which will use primary materials that require energy to obtain and produce. Energy efficiency estimates will need to include the amount of energy needed for this transition. Furthermore, if the economic savings produced by increased efficiencies are allocated to consumers in a highly commoditized economy, they will likely spend those savings on more commodities. The portion of the savings retained as profits are likely to go into new investments to produce more goods and services. The net result will

be continued and perhaps accelerated economic growth. If carried out globally the end result of even a tenfold increase in energy efficiency will lead
to massive worldwide increase in energy and materials use.

A strategy of deliberate decommoditization, replacing material and
energy with knowledge, mutual aid, and community, must therefore accompany any strategy of efficiency if the result is not going to be just increased
consumption of energy and materials. Yes, the commoditized economy must
be made more efficient and yes, it will be exciting and rewarding to accomplish this task. But the commoditized economy must also be made smaller,
and global development policy must turn away from the goal of growing
the commoditized economy toward a new goal, improvements in quality
of life with less energy and materials. With a decommoditization strategy
to accompany an energy efficiency policy, the money saved would be allocated toward community building, health maintenance and disease prevention, hands-on care for people in need, agricultural skills-building, land
conservation, liberal education, and the whole array of public goods presently marginalized by commoditization. A portion of the savings of the
efficiency and information revolutions must be effectively captured to serve
the public interest. The only way for that to occur is for the public to gain
or regain control over the institutions of governance and for those institutions to have the capacity to obtain and direct resources toward public goods
with little or no commodity potential.

Developing the political will to decommoditize will be more difficult
than achieving substantial energy efficiency gains, because decommoditization will have no obvious business allies. But it will also be a more popular
challenge, because it will cut across many different social movements that
have not previously clearly understood the connection between their goals
and decommoditization. A decommoditization strategy will mean having to
sever the close alliance between business and government that has created
the policies of commoditization. It will require removing money from politics
as much as possible. This creates the first of many possible political alliances
between disparate social movements, environmentalists, and the advocates
for good, responsive, publicly accountable government. Seeing that the vast
majority of the world's population favors good government and a healthy
environment, the popular base clearly exists for a social and political movement in support of the economy of care and connection.

According to Michael Walzer in his influential book *Spheres of Justice*
(1983), the very meaning of tyranny is when those who succeed in one
sphere of social life (such as money and commodities) deliberately and
unjustly wield the power and influence gained there to dominate in another
sphere of social life (such as political influence). The current role of some
multinational corporations has been well documented.[29] Liberty, according
to this formulation, requires active defense of the boundaries between the
spheres. But commoditization, which also leads to domination by one
sphere (money) over all others, is a "natural" outcome of system dynamics,
and the methods of defending against it are not nearly so clear as those in

the struggle against a tyrant who can be named and constrained or removed from power. Nonetheless, the movement for liberation from oppression involves much the same institutional framework that Walzer prescribes for the defense of liberty:

> a strong welfare state run, in part at least, by local and amateur public officials; a constrained market; an open and demystified civil service; independent public schools; the sharing of hard work and free time; the protection of religious and familial life; a system of public honoring and dishonoring free from all considerations of rank or class; workers' control of companies and factories; a politics of parties, movements, meetings and public debate.[30]

Such an institutional framework would need to consider the way economies evolve through selection pressures and deliberately design public policy instruments that provide countervailing forces to the pressures of commoditization. In addition to reinvigorating presently impoverished spheres of communal life, this would have the additional benefit of reducing the amount of materials and energy needed per unit of social welfare.[31]

This movement gains moral clout from the general understanding that there are many things of value that can't be bought and sold, and that it is the responsibility of society and its public institutions to protect these values against the corrosive effects of money. This common wisdom is given explicit meaning in the work of ecological economists, who demonstrate that the Earth provides the ultimate raw materials in the air we breathe, the forests that clean the air, the water that sustains us, and the sunlight that powers all work. These goods have not and cannot be adequately priced or fairly represented in private markets. The political implications of this are clear: society and its institutions must allocate resources for the protection of that which cannot be readily privatized. We cannot assume that private individuals, protecting their private property, will adequately be able to protect public property and public goods. This simple logic has enormous political consequences, because it calls forth the need for a greater role for society and its institutions to intervene in private markets for the public good. Advocates for the rights of women note that a large percentage of the work traditionally carried out by women, the work essential to the economy of caring and connection, is not compensated by private markets. The over-reliance on markets to allocate the wealth produced in the economy, therefore, is fundamentally unfair to women and a major reason for the inequality between the sexes. Ready to join the advocates for good government, environmental protection, and women's equality are the vast majority of the religions of the world who value exactly that which cannot be priced, who have a long and powerful tradition of speaking out about the corrupting influence of money in public life.

What is common to all these arguments about the need to properly value that which is not fairly priced in the free market is what could be called a politics of fair compensation. This is an idea around which much of the world can be mobilized. People understand notions of basic fairness and just rewards. In the popular understanding of justice there is an appreciation as well of the rightness of proper compensation for the victims and heirs of past wrongs. As we learn more and more of our history, we understand that the wealth of the North derived in large part from a long and tragic story of the exploitation of the labor, land, and resources of the South. If free markets left on their own continue this legacy of exploitation, then fairness and justice require that global civil society intervene through its institutions, its global trade agreements, its United Nations, and its international financial institutions to correct the inequality and compensate those from whom much was taken and little returned. A politics of just compensation would insist that enough of the wealth produced in the world be directed toward sustainable development of the poor so that all can have access to the basics of food, sanitation, health care, and education.

The great tragedy of commoditization is the way that development has been distorted. It is not possible to recover all the lost development opportunities we have foregone these many years. Who knows the varieties of crops Incan scientists could have produced, what brilliance of forestry the Iroquois could have managed given the time and resources. Our development path has been stunted and narrowed by the forces of commoditization and we have paid dearly, some more than others, all of us to a great degree. But we can start now. The opportunities are immense if we begin to invest heavily in sustainable development.

All around the world people are inventing and rediscovering models of development capable of delivering a high level of satisfaction of basic needs without over-exploiting the environment. These models are based as much as possible on local community and neighborhood self-sufficiency, small-scale and ecologically designed sanitation and water-delivery systems, small farms using a creative melding of modern technology and ancient ecological wisdom, appropriate and locally suited and modified technology and renewable energy. The price of spreading, supporting, and celebrating this model of development is small in dollar terms but enormously challenging politically. It will mean fundamental reforms, including land reform and democratization in favor of the poor. It will mean a sound strategy of decommoditization in the global economy, which in turn will require international institutions capable of raising and distributing sufficient amounts of capital to make a difference.

Just as the key to unleashing creative energies for commercial development lay in the invention and application of stimulative legal and political structures at the appropriate level of government, the key to unleashing noncommodity energies of caring and connection is also legal and political reform at the appropriate level. In the present economy this means having to regulate simultaneously at the global, national, and local levels.

Commercial law is now evolving into a global legal framework designed to unleash commercial energies worldwide by minimizing the capacity of states to restrict access to markets. As a result, commoditization pressures are expanding into all potential niches worldwide. Since the legal and political actors unleashing these forces operate at the global level, countervailing pressures must also operate globally. There must be legal and political reforms in the global trade and finance regimes. But since noncommodity solutions to human needs and wants are inherently local, the effects of these countervailing forces must be felt at the local level. New legal and political capacity to stimulate investment in community-based, less commoditized satisfactions for human needs and wants must evolve and go to the level nearest to the people with those needs and wants. There have been several efforts to describe the emergence of global civil society as a precursor to emerging governance capacity that can act with some affect and authority at both the global and local levels.[32] At the same time, nations must invent new legal frameworks that allow localities to innovate economically and that protect them from the colonizing impulses of global forces and actors.

One of the most compelling hopes for the future is the new leadership in Third World development that is emerging throughout the world. These leaders understand the history of colonialism and its effects; they have an understanding of ecology as well as a deep connection with the cycles and potentials of their own land. They are grounded in the values of traditional cultures and open to creative and locally appropriate technological innovations. In every part of the world thousands of these leaders are emerging, putting together the pieces of a strategy for Third World sustainable development. What is emerging is the beginnings of a powerful global social movement. This movement is being simultaneously organized at the local, regional, and global levels. It is a movement for justice, local self-reliance and development, international equality, and peace. If successful it will create the social conditions for the emergence of a new set of political institutions capable of acting in the global economy for the common good. It will create a new global political order and a sound and fair balance between the economy of commodities, which produces the world's wealth, and the economy of caring and connection, which supports the world's well being. The institutions of governance that will emerge will coexist at the local, bioregional, regional, and global levels, with problems handled at the lowest possible level, moving up in the structure only as the consequences — such as global warming, global capital movements, and global civil compacts — rise.

In the early part of the next century this movement will flourish in exciting and powerful ways. The leadership will come from people of the former colonial world, the indigenous people, women, particularly women of color, the small farmers, and traditional peoples. Their struggles to gain the political capacity to challenge entrenched systems of wealth and power will inspire millions of people, especially the young, in the wealthy enclaves of the world. Like youth movements of the past, these will gain alliances among the older generations, particularly the "baby boomers" of North

America as they head toward the later parts of their lives and begin again to focus on the things of this world that truly matter.[33] This will be an exciting time. The music, art, and celebrations of the Third World will be adopted and adapted by youth everywhere, as it already is in the popular music of the young. As the movement for sustainable development and justice grows in different parts of the world, their struggles and successes will inspire each other. The picture of determined people taking charge of their own destiny will be beamed throughout the world via TV and the Internet. Instantaneous communication will make it difficult for these movements to be suppressed and their leaders disappeared.

The objectives of this movement will be the protection of the health and integrity of the Earth's ecosystems, its diverse forms of life, and the elimination from the Earth of human exploitation, poverty, and misery — the two dreams that have been linked now irreversibly in the concept of sustainable development. As we have seen, development cannot be made sustainable without counteracting the pressures of commoditization. Sharing and improving an understanding of commoditization and its effects on society and the environment has been my goal in writing this book. My hope is that by understanding how commoditization works and how different policies and institutions can be designed to counterbalance it, the movement for caring and connection can be better situated to act strategically and effectively.

In the end we must each find proper balance in our own lives between our global dreams and our local connections. We can each direct our personal time, attention, and resources to make certain our individual and local community economy of caring and connection prospers. Where the institutions of governance fail us, we must join with others to reclaim them, reform them, and make them the tools by which we act for the common good. The laws of nature that determine the structure and design of our world are laws we cannot avoid. We can learn from them and learn how to live well within them, but we ignore them at our peril. The laws that determine the structure of our economy, on the other hand, are partly those of nature and partly of human design. What humans design we can change purposely and intelligently using the best information we have. The information we now have is that our economy privileges commodities over all else and thereby distorts our way of life so that the things that really matter — our connection and caring for each other and the Earth — are forgotten and ignored. We have the capacity to change this by making the economy accountable to governance, and governance accountable to citizens.

Notes

1. Wackernagel, M., Onisto, L., Callejas Linares, A., Lopez Falfan, I.S., Mendez Garcia, J., Suarez Guerrero, A.I., Suarez Guerrero, Ma. G., *Ecological Footprints of Nations, Report of the Centre for Sustainability Studies,* Universidad Anahuac de Xalapa, Mexico and The Earth Council, Costa Rica, 1997.

2. Schnaiberg, A., *The Environment from Surplus to Scarcity*, Oxford University Press, New York, 1980, p. 247.

3. Diaz-Briquets, S. and Pérez-Lopez, J., Socialism and environmental degradation, Presented at the Meeting of the Association for the Study of the Cuban Economy, Coral Gables, FL, August 6–8, 1998.

4. LaPorte Vest, A., Voluntary simplicity, *The Liguorian, 87* (3), 1999, pp. 10–13.

5. The covenant, signed by 99 countries, recognizes "the right of everyone to an adequate standard of living for himself and his family, including adequate food, clothing and housing, and to the continuous improvement of living conditions." See Brownlie, I., *Basic Documents of Human Rights*, 3rd ed., Clarendon Press, Oxford University Press, New York, 1992, p. 118.

6. Gersh, J., Capitalism goes green? Commonsense eco-fixes in the economic toolbox, *The Amicus Journal*, 21(1), Spring 1999, pp. 37–41.

7. McDonough, W. and Braungart, M., The NEXT industrial revolution, *Atlantic Monthly*, October 1998.

8. I first heard Jeremy Rifkin speak about the need to change our goals from efficiency to "sufficiency" at a conference in Washington in 1992 at the Second Global Structures Convocation on Environment and Development.

9. Ayres, R.U., Toward a zero-emmissions economy, *Environmental Science & Technology*, 32(15), 366A, August 1, 1998, p. 367A.

10. See Gates, J., *The Ownership Solution, Toward a Shared Capitalism for the 21st Century*, Perseus Books, Reading, MA, 1998.

11. Von Weizsacker, E., Ulrich, E., Lovins, A.B., and Lovins, L.H., *Facto Four: Doubling Wealth — Halving Resource Use*, Earthscan, London, 1997. Manshard, W., Eco-restructuring: the biophysical basis, in *Eco-Restructuring: Implications for Sustainable Development*, Ayres, R.U. (Ed.), United Nations University Press, Tokyo, 1998, p. 67. Schmidt-Bleek, F., MIPS — a universal ecological measurement, *Fresenius Ecological Bulletin*, 1, 306, 1992; also in *Fresenius Ecological Bulletin*, 2, 8, 1993, a special edition on the Material Inputs per Unit of Service [MIPS] project of the Wuppertal Institut für Klima, Umwelt, und Energie, Wuppertal, Germany, 1993.

12. Ayres, R.U., Eco-restructuring: the transition, in *Eco-Restructuring: Implications for Sustainable Development*, Ayres, R.U. (Ed.), United Nations University Press, Tokyo, 1998, p. 33.

13. Duchin, F., Eco-restructuring the macro economy, in *Eco-Restructuring: Implications for Sustainable Development*, Ayres, R.U. (Ed.), United Nations University Press, Tokyo, 1998, p. 266.

14. Douthwaite, R., *The Growth Illusion*, Council Oak Publishing, Tulsa, OK, 1993, p. 93.

15. United Nations Development Program, *United Nations Human Development Report 1998*, Oxford University Press, New York, 1998.

16. Reich, R., *The Work of Nations: Preparing Ourselves for 21st Century Capitalism*, Alfred A. Knopf, New York, 1991, p. 8.

17. Henwood, D. The free flow of money, *Third World Resurgence*, Issue No. 99, 1/98, pp. 17–23.

18. Al Haq, M. (Ed.), *The Tobin Tax: Coping with Financial Volatility*, Oxford University Press, New York, 1996.

19. See Hammond, M.J. and Dunkiel, B., The logic of environmental tax reform in the United States, *Ecological Economics Bulletin*, 3(2), 1998, 25–27.

20. Rees, W., Reducing the ecological footprint of consumption, in *The Business of Consumption: Environmental Ethics and the Global Economy*, Westra, L. and Werhane, P. H. (Eds.), Rowman & Littlefield, Lanham, MD, 1998, pp. 113–130.

21. Brown, L., Flavin, C. and Postel, S., *Saving the Planet: How to Shape an Environmentally Sustainable Global Economy*, Worldwatch Institutute, Washington, D.C., 1991; see also Roodman, D.M., *The Natural Wealth of Nations*, W.W. Norton, New York, 1998.

22. Alliance to Save Energy (ASE), *Federal Energy Subsidies: Energy, Environmental and Fiscal Impacts*, The Alliance to Save Energy, Washington, D.C., 1993.

23. Worldwatch Institute, *State of the World*, Worldwatch Institute, Washington, D.C., 1989.

24. Roodman (1998), p. 63.

25. Sutton, P., A speculative sketch of a successful ecologically-sustainable economy, Green Innovations Inc., 1998, Website http://www.peg.apc.org/~psutton/sectors.htm, 5/12/99.

26. Witzeling, R., *People, Not Profit, the Story of the Credit Union Movement*. Kendall/Hunt, Dubuque, IA, 1993. See also Isbister, J., *Thin Cats: The Community Development Credit Union Movement in the United States*, Center for Cooperative, University of California, 1994.

27. Ayres, E. (Ed.), *Worldwatch*, Worldwatch Institute, Washington, D.C., 1999.

28. Lovins, A.B. and Lovins, L.H., *Climate: Making Sense* and *Making Money*, Report of the Rocky Mountain Institute, Old Snowmass, CO, 11/13/97.

29. Kurlansky, M., Oil, toil and tyranny, *Audubon*, 128, 1996.

30. Walzer, M., *Spheres of Justice: A Defense of Pluralism and Equality*, Basic Books, New York, 1983, p. 318.

31. Daly, H.E. and Cobb J.B., Jr., *For the Common Good: Redirecting the Economy toward Community, the Environment and a Sustainable Future*, Beacon Press, Boston, 1989.

32. Walzer (1993); Princen, T. and Finger, M. [w/contributions of Manno, J. P. and Clark, M. L.], *Environmental NGOs in the World Politics: Linking the Local and the Global*, Routledge, London, 1994; Lipschutz, R. D., *Global Civil Society and Global Environmental Governance: The Politics of Nature from Place to Planet*, State University of New York Press, Albany, 1996; Wapner, P.K., *Environmental Activism and World Civic Politics*, State University of New York Press, Albany, 1996.

33. Roszak, T., *America the Wise, the Longevity Revolution and the True Wealth of Nations*, Houghton Mifflin, Boston, 1998.

References

Adams, D., *Renewable Resource Policy*, Island Press, Washington, D.C., 1993.

Aguilera-Klink, F., Some notes on the misuse of classic writings in economics on the subject of common property, *Ecological Economics*, 9, 1994, 221–228.

Al Haq, M. (Ed.), *The Tobin Tax: Coping with Financial Volatility*, Oxford University Press, New York, 1996.

Alliance for Responsible Trade & Common Frontiers, *Alternatives for the Americas, Building a People's Hemispheric Agreement*, The Canadian Centre for Policy Alternatives & Common Frontiers, Ottawa, 1999.

Alliance to Save Energy (ASE), *Federal Energy Subsidies: Energy, Environmental and Fiscal Impacts,* The Alliance to Save Energy, Washington, D.C., 1993.

American Automobile Manufacturers Association, *Motor Vehicle Facts & Figures 1997*, American Automobile Manufacturers Association, Washington, D.C., 1997, http://www/aama.com.

Anderson, M.D., The life cycle of alternative agricultural research, *American Journal of Alternative Agriculture*, 10(1), 1995.

Arthur, W.B., *Increasing Returns and Path Dependence in the Economy*, University of Michigan Press, Ann Arbor, 1994.

Ausubel, J. H., Can technology spare the Earth? *American Scientist*, 84, 166, 1996.

Ayres, E. (Ed.), *Worldwatch*, Worldwatch Institute, Washington, D.C., 1999.

Ayres, R.U., Industrial metabolism, theory and policy, *Industrial Metabolism. Restructuring for Sustainable Development*, Ayres, R.U., and Simonis, U.E. (Eds.), United Nations University Press, Tokyo, 1994, Chapter 1.

Ayres, R.U., Eco-restructuring: the transition, in *Eco-restructuring: Implications for Sustainable Development*, Ayres, R.U. (Ed.), United Nations University Press, Tokyo, 1998, 1.

Ayres, R.U., Toward a zero-emmissions economy, *Environmental Science & Technology*, 32(15), 366A, August 1, 1998.

Azfar, K., *Asian Drama Revisited*, Royal Book Company, Karachi, 1992.

Baum, G., *Karl Polanyi on Ethics & Economics*, McGill-Queen's University Press, Montreal, 1996.

Bellush, J. and Hausknecht, M. (Eds.), *Urban Renewal: People, Politics, and Planning*, Doubleday, New York, 1967.

Berkes, F., The benefits of the commons, *Nature*, 340, July 13, 1989, 91.

Boulding, K. E., *Ecodynamics: A New Theory of Societal Evolution*, Sage Publications, Beverly Hills, CA, 1978.

Brown, L., Flaven, C. and Postel, S., *Saving the Planet: How to Shape an Environmentally Sustainable Global Economy,* Worldwatch Institute, Washington, D.C., 1991.

Brown, L.R. and Kane, H., Full House: *Reassessing the Earth's Population Carrying Capacity,* W. W. Norton, New York, 1994.

Brown, L.R., Gardner, G., and Halwell, B., *Beyond Malthus: Sixteen Dimensions of the Population Problem, Worldwatch Paper 143,* Worldwatch Institute: Washington, D.C., 1998.

Brown, L.R., Flavin, C., and French, H., *State of the World 1999, A Worldwatch Institute Report,* W. W. Norton, New York, 1999.

Brownlie, I. (Ed.), *Basic Documents on Human Rights,* 3rd ed., Clarendon Press (Oxford University Press), New York, 1992.

Bureau of the Census, *Survey of Environmental Products and Services,* United States Environmental Protection Agency, Washington, D.C., February 1998.

Carrithers, M., *Why Humans Have Cultures, Explaining Anthropology and Human Diversity,* Oxford University Press, New York, 1992.

Catton, W. R., *Overshoot,* University of Illinois Press, Urbana, IL, 1980.

Chandler, A., *The Visible Hand,* The Belknap Press, Cambridge, MA, 1977.

CIS Statistical Universe, No. 154, National Health Expenditures, by object: 1980 to 1995, http://web.lexis-nexis.com/statuniv/attachment/attachment.gif ?_butinfo=, 4/12/99.

CIS Statistical Universe, No. 173 Health Maintenance Organizations (HMOs): 1980 to 1995, http://web.lexis-nexis.com/statuniv/attachment/attachment.gif ?_butinfo=, 4/12/99.

Cobb, C., Halstead, T., and Rowe, J., If the GDP is up, why is America down? *The Atlantic Monthly,* 276(4), 59, 1995.

Cohen, J.E., Population growth and Earth's human carrying capacity, *Science,* 269, 341, 1995.

Commons, J., *Legal Foundations of Capitalism,* The University of Wisconsin Press, Madison, 1968.

Conservation Law Foundation, *Road Kill: How Solo Driving Runs Down the Economy. A Conservation Law Foundation Report, May 1994.,* CLF, Boston, 1994.

Costanza, R., Ecological economics, reintegrating the study of humans and nature, *Ecological Applications,* 6, 978, 1996.

Costanza, R., Cumberland, J., Daly, H., Goodland, R., and Norgaard, R., *An Introduction to Ecological Economics,* St. Lucie Press, Boca Raton, FL, 1997.

Cottrell, F., *Energy and Society,* McGraw-Hill, New York (reprinted by Greenwood Press, Westport, CT), 1955.

Cronon, W., *Changes in the Land,* Hill and Wang, New York, 1990.

Cross, G., *Kid's Stuff, Toys and the Changing World of American Childhood,* Harvard University Press, Cambridge, MA, 1997.

Cross, J.G. and Guyer, M.J., *Social Traps,* University of Michigan Press, Ann Arbor, MI, 1980.

Daily, G.C. and Ehrlich P., Socioeconomic equity, sustainability, and Earth's carrying capacity, *Ecological Applications,* 6, 991, 1996.

Daly, H.E., From empty world economics to full world economic: recognizing an historic turning point in economic development, in Goodlands, R., Daly, H. H., El Serafy, S., and Von Droste, B. (Eds.), *Environmentally Sustainable Economic Development: Building on Bruntland,* UNESCO, Paris, 1991.

Daly, H.E., Commentary on Nicholas Georgescu-Roegen's contributions to economics: an obituary essay, *Ecological Economics,* 13, 149, 1995.

Daly, H.E. and Cobb, J.B., Jr., *For the Common Good: Redirecting the Economy toward Community, the Environment and a Sustainable Future*, Beacon Press, Boston, 1989.

Daly, H.E. and Townsend, K.N., *Valuing the Earth: Economics, Ecology and Ethics*, MIT Press, Cambridge, MA, 1993.

Daly, H.E., On Nicholas Georgescu-Roegen's contributions to economics, *Ecological Economics*, 13, 1995, 149–54.

Daly, H.E., *Beyond Growth: The Economics of Sustainable Development*, Beacon Press, Boston, 1996.

Davis, G. R., Energy for planet Earth, in *Energy for Planet Earth*, Piel, J. (Ed.), W. H. Freeman, New York, 1991, 1.

DePinto, J., Manno, J., Milligan, M., Guzman, R., Honan, J., and Lopez Lemus, L., *Proposed Chlorine Sunsetting in the Great Lakes Basin: Policy Implications for New York State*, Donald W. Rennie Memorial Monograph Series, No. 11, State University of New York at Buffalo, NY, 1998.

Diamond, J., *Guns, Germs and Steel*, W. W. Norton, New York, 1997.

Diaz-Briquets, S. and Pérez-Lopez, J., Socialism and environmental degradation, presented at the Meeting of the Association for the Study of the Cuban Economy, Coral Gables, FL, August 6–8, 1998.

Dickson, D., *Alternative Technology and the Politics of Technical Change*, Wm. Collins & Sons, Glasgow, 1974.

Dolan, M. and Abramson, R., Expectations for summit come down to earth, *LA Times*, home edition, Part A, June 1, 1992.

Douthwaite, R., *The Growth Illusion*, Council Oak Publishing, Tulsa, OK, 1993.

Duchin, F., Eco-restructuring the macro economy, in *Eco-restructuring: Implications for Sustainable Development*, Ayres, R.U. (Ed.), United Nations University Press, Tokyo, 1998, 259.

Editors of the Ecologist, *Whose Common Future? Reclaiming the Commons: the Ecologist*, New Society Publishers, Philadephia, 1993.

Edwards, R., Tomorrow's bitter harvest, *New Scientist*, 151(2043), August 17, 1996, 14.

Ehrlich, P.R. and Wilson, E.O., Biodiversity studies: science and policy, *Science*, 253, 758, 1991.

Ehrlich, P.R., Ehrlich, A.H., and Daily, G.C., *The Stork and the Plow*, G.P. Putnam, New York, 1995.

Ehrlich, P.R., Ehrlich, A.H., and Holdren, J.P., Availability, entropy, and the laws of thermodynamics, in *Valuing the Earth: Economics, Ecology, Ethics*, Daly, H.E. and Townsend, K.N. (Eds.), MIT Press, Cambridge, MA, 1993, 69.

Elton, C., *The Pattern of Animal Communities*, Methuen, London, 1966.

Environment Canada, *Canada's Green Plan for A Healthy Environment*, Government of Canada, Ottawa, 1990.

Erekson, O.H., Loucks, O., and Strafford, N., The context of sustainability, *Sustainability Perspectives in Business and Resources*, Loucks, O., Erekson, H., Bol, J. W., Grant, N., Gorman, R., Johnson, P. and Krehbiel, T. (Eds.), CRC/St. Lucie Press, Delray Beach, FL, 1998.

Farquhar, J.W., Case for dissemination research in health promotion and disease prevention, *Canadian Journal of Public Health*, 87(2), S44, Nov–Dec, 1996.

Folke, C., Jansson, A., Larson, J., and Costanza, R., Ecosystem appropriation by cities, *Ambio*, 26, 167, 1997.

Freund, P. and Martin, G., *The Ecology of the Automobile*, Black Rose Books, New York, 1993.

Fried, M., Grieving for a lost home, *The Urban Condition*, Duhl, L.J. (Ed.), Basic Books, New York, 1963.

Fuglie, K., Ballenger, N., Day, K., Klotz, C., Ollinger, M., Reilly, J., Vasavada, U., and Yee, J., *Agricultural Research and Development: Public and Private Investments under Alternative Markets and Institutions*, U.S. Department of Agriculture, Agricultural Economic Report Number 735, Washington, D.C., May 1996.

Gallopin, G. C., Eco-restructuring tropical land use, in *Eco-restructuring: Implications for Sustainable Development*, Ayres, R.U. (Ed.), United Nations University Press, Tokyo, 1998.

Gates, J., *The Ownership Solution, Toward a Capitalism for the 21st Century*, Perseus Books, New York, 1998.

Georgescu-Roegen, N., *The Entropy Law and the Economic Process*, Harvard University Press, Cambridge, MA, 1971.

Gersh, J., Capitalism goes green? Commonsense eco-fixes in the economic toolbox, *The Amicus Journal*, 21(1), Spring 1999.

Goldemberg, J., Energy needs in developing countries and sustainability, *Science*, 269, 1058, August 25, 1995.

Goldemberg, J., Johansson, T. B., Reddy, A. K. N. and Williams, R. H., Basic needs and much more with one kilowatt per capita, *Ambio*, 14(4–5), 190, 1985.

Golley, F., *A History of the Ecosystem Concept in Ecology*, Yale University Press, New Haven, CT, 1993.

Goodland, R., Daly, H.H., El Serafy, S., and Von Droste (Eds.), *Environmentally Sustainable Economic Development: Building on Bruntland*, UNESCO, Paris, 1991.

Goodland, R. and Daly, H.E., Environmental sustainability: universal and non-negotiable, *Ecological Applications*, 6, 1002, 1996.

Gore, A., *Earth in Balance: Ecology and the Human Spirit*, Houghton Mifflin, Boston, 1992.

Gorz, A., *Ecology as Politics*, South End, Boston, 1980.

Gould, K. A., Schnaiberg, A., and Weinberg, A. S., *Local Environmental Struggles: Citizen Activism in the Treadmill of Production*, Cambridge University Press, New York, 1996.

Hall, C.A.S., Cleveland, C. J., and Kaufmann, R., *Energy and Resource Quality: The Ecology of the Economic Process*, University Press of Colorado, Niwot, CO, 1992.

Hall, C.A.S. and Hall, M.H.P., Issues in global agricultural and environmental sustainability: the efficiency of land and energy use in tropical economies and agriculture, *Agriculture, Ecosystems and Environment*, 46, 1, 1993.

Hall, C.A.S., Pontius, G., Ko, J.-Y., and Coleman, L., The environmental impact of having a baby in the United States, *Population and Environment*, 15, 505, 1994.

Hall, C.A.S., Economic development or developing economics: what are our priorities? *Ecosystem Rehabilitation*, Wali, M.K. (Ed.), SPB Academic Publishing, The Hague, 1992, 101.

Hall, C.A.S., The environmental impact of U.S. babies, *Earth Island Journal*, 10(4), 19, Fall 1995.

Hall, C.A.S., Ricardo lives: the inverse relation of resource exploitation intensity to efficiency, and its relation to sustainable development (unpublished paper), 1998.

Hammond, M.J. and Dunkiel, B., The logic of environmental tax reform in the United States, *Ecological Economics Bulletin* 3 (2), 1998, 25–27.

Hannon, B., Energy conservation and the consumer, *Science*, 189(4197), 95, July 11, 1975.

Hardin, G., The tragedy of the commons, *Science*, 162, 1243, 1968.

Harwood, R.R., History of sustainable agriculture: U.S. and international perspective, *Sustainable Agricultural Systems*, Edwards, C.A., Lal, R., Madden, P., Miller, R.H. and House, G. (Eds.), Soil and Water Conservation Society, Ankeny, IA, 1990.

Hecht, S. and Cockburn, A., *The Fate of the Forest*, HarperPerennial, New York, 1990.

Hecht, S., Indigenous soil management, *Let Farmers Judge*, Hiemstra, W., Reijntjes, C., and van der Werf, E. (Eds.), Intermediate Technology Publications, London, 1992.

Henwood, D., The free flow of money, *Third World Resurgence*, Issue No. 99, 1/98, 17–23.

Hinterberger, F., Biological, cultural, and economic evolution and the economy/ecology relationship, in *Toward Sustainable Development: Concepts, Methods and Policy*, van den Bergh, J.C.J.M. and van der Straaten, J. (Eds.), Island Press, Washington, D.C., 1994, 57.

Hobbes, C.F. and Kiggundu, M. N., *A Global Analysis of Life Expectancy and Infant Mortality*, Carleton University Press, Ottawa, 1992.

Holling, C.S., What barriers? what bridges? *Barriers & Bridges to the Renewal of Ecosystems and Institutions*, Gunderson, L.H., Holling, C. S., and Light, S. (Eds.), Columbia University Press, New York, 1995.

Holmberg, J.K., Robért, H. and Eriksson, K. E., Socio-ecological principles for a sustainable society: scientific background and Swedish experience, in *Socio-Ecological Principles and Indicators for Sustainability*, Holmberg, J. (Ed.), Institute of Physical Resource Theory, Göteborg, Sweden, 1995.

Hook, W., Counting on cars, counting out people, *Institute for Transportation Development Policy Paper*, New York, Winter 1994.

Hornborg, A., Ecosystems and world systems: accumulation as an ecological process, *Journal of World-Systems Research*, 4, 169, 1998.

Horton, D.E., Lessons from the Mantaro Valley project, Peru, in *Let Farmers Judge: Experiences in Assessing the Sustainability of Agriculture*, Hiemstra, W., Reijntjes, C., and van der Werf, E. (Eds.), Intermediate Technology Publications, London, 1992.

Hornborg, A., Ecsosystems and world systems: accumulation as an ecological process, *Journal of World-Systems Research*, 4(2), Fall 1998, 169–177.

Hueting, R., Three persistent myths in the environmental debate, *Ecological Economics*, 18, 81, 1996.

Hurst, J.W., *Law and the Conditions of Freedom: in the 19th Century United States*, University of Wisconsin Press, Madison, 1956.

Hutchinson, G.E., Population studies: animal ecology and demography, Cold Spring Harbor Symposia on Quantitative Biology, 22, 415, 1957.

Hynes, P.H., *Taking Population out of the Equation: Reformulating I=PAT*, Institute on Women and Technology, North Amherst, MA, 1993.

Idris, S.M.M., The message from Calicut — 500 years after Vasco da Gama, *Third World Resurgence*, 96, 1998.

Illich, I., *Tools for Conviviality*, Harper & Row, New York, 1973.

IIED, Citizen Action to Lighten Britain's Ecological Footprint, report prepared by the International Institute for Environment and Development for the U.K. Department of the Environment, International Institute for Environment and Development, London, 1995.

Interface Flooring Systems, Inc., Website at: http //216.1.140.49/us, 5/4/99.

Isbister, J., *Thin Cats: The Community Development Credit Union Movement in the United States*, Center for Cooperative, University of California, 1994.

Jevons, W.S., *The Coal Question: an Enquiry Concerning the Progress of the Nation, and the Probable Exhaustion of our Coal Mines*, MacMillan, London, 1865.

Jockinsen, M. and Knobloch, U., Making the hidden visible: the importance of caring activities and their principles for any economy, *Journal of Ecological Economics,* 20(2), 1997.

Kaplan, R., The coming anarchy, *Atlantic Monthly,* 273(2), 44–76.

Kauffman, S., *At Home in the Universe: The Search for the Laws of Self-Organization and Complexity,* Oxford University Press, New York, 1995.

Kay, J. H., *Asphalt Nation,* Crown Books, New York, 1997.

Keynes, J. M., *The General Theory of Employment, Interest and Money,* Harcourt, Brace, New York, 1936.

Klotz, C., Fuglie, K. and Pray, C., *Private-sector Agricultural Research Expenditures in the United States, 1960–92,* U.S. Department of Agriculture, Natural Resources and Environment Division, Economic Research Service Staff Paper No. AGES9525, October 1995.

Ko, J.-Y., Hall, C.A.S., and Lopez Lemus, L.G., Resource use rates and efficiency as indicators of regional sustainability: an examination of five countries, *Environmental Monitoring and Assessment,* 51, 571, 1998.

Kohn, A., *No Contest: The Case against Competition,* revised ed., Houghton Mifflin, New York, 1992.

Korten, D.C., Sustainable development, *World Policy Journal,* 9(1), 157, 1992.

Kurlansky, M., Oil, toil and tyranny, *Audubon,* 128, 1996.

Kunstler, J.H., *The Geography of Nowhere: The Rise and Decline of America's Man-Made Landscape,* Simon & Schuster, New York, 1993.

Kuznets, S.S., How to judge quality, *The New Republic,* 147, October 20, 1962.

Kwa, C., Representations of nature mediating between ecology and science policy: the case of the International Biological Programme, *Social Studies of Science,* 17, 1987, 413–442.

Lamptey, P. and Cates, W., An ounce of prevention worth a million lives, *Network,* 17(2), Family Health International, Winter 1977.

Landsberg, H. H., Materials: some recent trends and issues, *Science,* February 20, 1976, 637.

LaPorte Vest, A., Voluntary simplicity, *The Liguorian,* 87 (3), 1999, 10–13.

Lappé, M. and Bailey, B., *Against the Grain: Biotechnology and the Corporate Takeover of Your Food,* Common Courage Press, Monroe, ME, 1998.

LaRosa, J.H. and Kiefhaber, A., Cost analysis and workplace health promotion programs, *Occupational Health Nursing,* 33, 234, 1985.

Leopold, A., *Sand County Almanac,* Oxford University Press, New York, 1968.

Lietaer, B., From the real economy to the speculative, *Third World Economics,* 172, November 1997, 15.

Lilliston, L. and Cummins, R., Organic versus 'organic': the corruption of a label, *The Ecologist,* July/August 1998.

Lipschutz, R.D., *Global Civil Society and Global Environmental Governance: The Politics of Nature from Place to Planet,* State University of New York Press, Albany, 1996.

Lipson, M., *Searching for the "O-Word,"* Organic Farming Research Foundation, Santa Cruz, CA, 1997.

Lotka, A. J., Contributions to the energetics of evolution, *Proceedings of the National Academy of Sciences,* 8, 147, 1922.

Lovins, A. and Browning, W.D., Negawatts for Buildings, *Urban Land,* 51(7), July, 1992

Lovins, A. and Lovins, L.H., *Climate: Making Sense and Making Money,* Rocky Mountain Institute, Old Snowmass, CO, 1997.

Lovins, A. and Lovins, L.H., Reinventing the wheels, *The Atlantic Monthly*, Boston, January 1995.

Lubchenko, J., Entering the century of the environment: a new social contract for science, *Science*, 279, January 23, 1998.

Magaziner, I. and Reich, R.B., *Minding America's Business*, Vintage, New York, 1982.

Manshard, W., Eco-restructuring: the biophysical basis, in *Ecorestructuring: Implications for Sustainable Development*, Ayres, R.U. (Ed.), United Nations University Press, Tokyo, 1998, 55.

Jessica Matthews of the World Resources Institute quoted in Ayres, R.U., Eco-restructuring: the transition, in Ayres, R.U. (Ed.), *Eco-restructuring: Implications for Sustainable Development*, United Nations University Press, Tokyo, 1998, 21.

Matson, P.A., Parton, W.J., Power, A.G., and Swift, M.J., Agricultural intensification and ecosystem properties, *Science*, 277, 504, July 25, 1997.

Max-Neef, M., Economic growth and quality of life: a threshold hypothesis, *Ecological Economics*, 15(2), 115, 1995.

Maynard Smith, J. and Price, G.R., The logic of animal conflict, *Nature*, 246 (5427), 15, November 2, 1973.

McDonough, W. and Braungart, M., The NEXT industrial revolution, *Atlantic Monthly*, October 1998.

McIntosh, R.P., *Ecology since 1900: History of American Ecology*, Arno Press, New York, 1977.

McKinney, M. and Schoch, R., *Environmental Science: Systems and Solutions*, Jones & Bartlett, Boston, 1998.

Meadows, D.H., Meadows, D.L., and Randers, J., *Beyond the Limits*, Chelsea Green, Post Mills, VT, 1992.

Meier, G.M., *Leading Issues in Economic Development*, 5th ed., Oxford University Press, New York, 1989.

Milani, B., What is green economics, Website of the Ecomaterials Project, Toronto: http://www.web.net/~bmilani/what.htm, 5/3/98.

Miller, G.T., Jr., *Living in the Environment: Principles, Connections, and Solutions*, Wadsworth, Belmont, 1996.

Miller, T., The Costs of Highway Crashes, The Urban Institute, Washington, D.C., October 1991.

Morris, C., *The Great Legal Philosophers: Selected Readings in Jurisprudence*, University of Pennsylvania Press, Philadelphia, 1978.

Morris, D., Future fibers, *Co-op America Quarterly*, Summer 1998, 7.

Moser, A., Ecological engineering and bioprocessing, in *Eco-restructuring: Implications for Sustainable Development*, Ayres, R.U. (Ed.), United Nations University Press, Tokyo, 1998, p. 77.

Munasinghe, M. and McNeely, J., Key concepts and terminology of sustainable development, *Defining and Measuring Sustainability: The Biogeophysical Foundations*, Distributed for the United Nations by the World Bank, Washington, D.C., 1995, chapter 2, 19.

Murdoch, W., World hunger and population, in *Agroecology*, Carroll, C. R., Vandermeer, J. H., and Rossett, P. M. (Eds.), McGraw-Hill, New York, 1990, 3.

Musson, A. E. (Ed.), *Science, Technology, and Economic Growth in the 18th Century*, Columbia University Press, New York, 1972.

Myers, N., *A Wealth of Wild Species*, Westview Press, Boulder, CO, 1983.

Myrdal, G., *Asian Drama: An Inquiry into the Poverty of Nations*, Twentieth Century Fund, New York, 1968.

Nadis, S. and MacKenzie, J., *Car Trouble*, Beacon Press, Boston, 1993.

National Organic Program Webpage, http://www.ams.usda.gov/nop/rule/summ1.htm, 2/17/99.

National Research Council, *U.S. Participation in the International Biological Program*, National Academy of Sciences, Washington, D.C., 1974.

National Research Council, *Sustainable Agriculture and the Environment in the Humid Tropics*, National Academy Press, Washington, D.C., 1993.

Nordhaus, W. and Tobin, J., Is growth obsolete? *Economic Growth, National Bureau of Economic Research General Series*, No. 96E, Columbia University Press, New York, 1972.

Norgaard, R., *Development Betrayed: The End of Progress and a Coevolutionary Revisioning of the Future*, Routledge, London, 1994.

North, D., *Structure and Change in Economic History*, W.W. Norton, New York, 1981.

North, D. and Thomas, R., *The Rise of the Western World: A New Economic History*, Cambridge University Press, Cambridge, 1973.

Odum, E.P., *Ecology and Our Endangered Life-Suppport Systems*, 2nd ed., Sinauer Associates, Inc., Sunderland, MA, 1993.

Odum, H.T., *Environmental Accounting: Energy and Decision Making*, John Wiley, New York, 1996.

Olson, D.W. and Jasinski, L., Keyboard efficiency, *Byte*, February 1986.

Ophuls, W., *Requiem for Modern Politics*, Westview, Boulder, CO, 1997.

Organic Farming Research Foundation, Frequently Asked Questions About Organic Farming, at http://www.ofrf.org. July 17, 1998.

Orr, D.W., *Ecological Literacy: Education and the Transition to a Postmodern World*, State University of New York Press, Ithaca, NY, 1992.

Ostrom, E., *Governing the Commons: The Evolution of Institutions for Collective Action*, Cambridge University Press, New York, 1990.

Pahlman, C., Lessons from the Mantaro Valley project, Peru, in *Let Farmers Judge*, Hiemstra, W.C., Reinjntjes, C., and van der Werf, E. (Eds.), Intermediate Technology Publications, London, 1992, 29.

Paoletti, M.G. and Pimentel, D., Genetic engineering in agriculture and the environment, *Bioscience*, 46 (9), 665, 1996.

Partridge, E., Reconstructing ecology, in *Global Ecological Integrity*, Pimentel, D. (Ed.), Island Press, Washington, D.C., 1999.

Pecora, N.O., *The Business of Children's Entertainment*, Guilford Press, New York, 1998.

Peet, J., *Energy and the Ecological Economics of Sustainability*, Island Press, Washington, D.C., 1992.

Pender, N., *Health Promotion in Nursing Practice*, Appleton & Lange, New York, 1996, 299.

PhRMA, 49 New Treatments Added to Nation's Medicine Chest; Companies to Invest $20.6 Billion in Research, January, 1998, Web publication http://www.phrma.org/charts/nda%5Fi97.html, 10/30/98.

Pimentel, D., Jackson, W., Bender, M., and Pickett, W., Perennial grains: an ecology of new crops, *Interdisciplinary Science Review*, 11, 42, 1986.

Pimentel, D., Stachow, U., Takacs, D.A., Brubaker, H.W., Dumas, A.R., Meaney, J.J., O'Neal, J.A.S., Onsi, D.E., and Corzilius, D.B., Conserving biological diversity in agricultural/forestry systems, *Bioscience*, 42(5), May 1992, 354.

Pimentel, D., Harvey, C., Resosudarmo, P., Sinclair, K., Kurtz, D., McNair, M., Crist, S., Spritz, L. Fitton, L., Saffouri, R. and Blair, R., Environmental and economic costs of soil erosion and conservation benefits, *Science*, 267, 111, 1995.

Pimentel, D. and Pimentel, M. (Eds.), *Food, Energy and Society,* University Press of Colorado, Niwot, 1996.

Pimentel, D., Environmental sustainability and integrity in the agriculture sector, *Global Ecological Integrity,* Pimentel, D. (Ed.), in press.

Polanyi, K., *The Great Transformation,* Beacon Hill, Boston, 1957.

Power, T.M., *The Economic Pursuit of Quality,* M.E. Sharpe, Armonk, 1988.

Princen, T. and Finger, M. [w/contributions of Manno, J.P. and Clark, M.L.], *Environmental NGOs in the World Politics: Linking the Local and the Global,* Routledge, London, 1994.

Ray, D., *Development Economics,* Princeton University Press, Princeton, NJ, 1998.

Redcliff, M., Reflections on the 'sustainable development' debate, *International Journal of Sustainable Development and World Ecology,* 3(21), 4, 1994.

Reddy, A.K.N and Goldemberg, J., Energy for the developing world, *Scientific American,* September 1990, 111.

Rees, W.E. and Wackernagel, M., Ecological footprints and appropriated carrying capacity: measuring the natural capital requirements of the human economy, in *Investing in Natural Capital: The Ecological Economics Approach to Sustainability,* Jansson, A.M., Hammer, M., Folke, C., and Costanza, R. (Eds.), Island Press, Washington, D.C., 1994.

Rees, W.E., More jobs, less damage: a framework for sustainability, growth and employment, *Alternatives,* 21, 24, 1995.

Rees, W.E., Taxing combustion and rehabilitating forests: achieving sustainability, growth and employment through energy policy, *Alternatives,* 21, 31, 1995.

Rees, W.E., Reducing the ecological footprint of consumption, *The Business of Consumption: Environmental Ethics and the Global Economy,* Westra, L. and Werhane, P. (Eds.), Rowman and Littlefield, Lanham, MD, 1998.

Rees, W.E., Patch disturbance, dissipative structure and ecological integrity: a "second law" synthesis, in *Global Ecological Integrity,* Pimentel, D. (Ed.), Island Press, Washington, D.C., in press.

Reich, R., *The Work of Nations: Preparing Ourselves for 21st Century Capitalism,* Alfred A. Knopf, New York, 1991.

Renner, M., *Jobs in a Sustainable Economy, WorldWatch Paper 104,* WorldWatch Institute, Washington, D.C., 1991.

Replogle, M., Improving Access for the Poor in Urban Areas, *Appropriate Technology,* 20(1),1993.

Robert, K.H., Educating a nation: the natural step, *In Context,* 28, 10, 1991.

Rogner, H.H., Global energy futures: the long-term perspective for eco-restructuring, in *Eco-restructuring: Implications for Sustainable Development,* Ayres, R.U. (Ed.), United Nations University Press, Tokyo, 1998, 149.

Rohatgi, P., Rohatgi, K., and Ayres, R., 1998. Materials futures: pollution prevention, recycling and improved functionality, in *Eco-restructuring: Implications for Sustainable Development,* Ayres, R.U. (Ed.), United Nations University Press, Tokyo, 1998, 109.

Roodman, D.M., *Paying the Piper: Subsidies, Politics and the Environment, WorldWatch Paper 13,* WorldWatch Institute, Washington, D.C., 1996.

Roodman, D.M., *The Natural Wealth of Nations,* W.W. Norton, New York, 1998.

Rosenberg, N. and Birdzell, L.E., Jr., *How the West Grew Rich: The Economic Transformation of the Industrial World,* Basic Books, New York, 1986.

Rostow, W.W., *The Stages of Economic Growth; a Non-Communist Manifesto,* University Press, Cambridge, 1971.

Roszak, T., *America the Wise, the Longevity Revolution and the True Wealth of Nations*, Houghton Mifflin, Boston, 1998

Sale, K., *Dwellers in the Land, the Bioregional Vision*, Sierra Club Books, San Francisco, 1985.

Sale, K., *The Conquest of Paradise*, Alfred A. Knopf, New York, 1990.

Sale, K., *Rebels against the Future*, Addison-Wesley, Reading, MA, 1995.

Sanchez, P. and Benites, J., Low-input cropping for acid soils of the humid tropics, in *Let Farmers Judge*, Hiemstra, W., Reijntjes, C., and van der Werf, E. (Eds.), Intermediate Technology Publications, London, 1992, 115.

Sapon-Shevin, M., *Because We Can Change the World*, Allyn and Bacon, Boston, 1999.

Schmidheiny, S., Food and agriculture, *Changing Course: A Global Business Perspective on Development and the Environment*, Schmidheiny, S. and the Business Council for Sustainable Development (Eds.), MIT Press, Cambridge, MA, 1992.

Schmidt-Bleek, F., MIPS — a universal ecological measurement, *Fresenius Ecological Bulletin*, 1, 306, 1992; also in *Fresenius Ecological Bulletin*, 2, 8, 1993, a special edition on the Material Inputs per Unit of Service [MIPS] project of the Wuppertal Institut für Klima, Umwelt, und Energie, Wuppertal, Germany, 1993.

Schnaiberg, A., *The Environment from Surplus to Scarcity*, Oxford University Press, New York, 1980.

Schnaiberg, A., *Environment and Society: The Enduring Conflict*, St. Martin's Press, New York, 1994.

Schultz, T., *Transforming Traditional Agriculture*, Yale University Press, New Haven, CT, 1964.

Schumacher, E. F., *Small is Beautiful*, Harper and Row, New York, 1973.

Smithsonian Institute Webpage, http://www.si.edu/biodiversity/aislynn1.htm, 5/7/99.

Stahel, W. and Jackson, T., Optimal utilization and durability, *Clean Production Strategies*, Jackson, T. (Ed.), Lewis Publishers [for the Stockholm Environment Institute], London, 1993.

Stevens, T., Playing to win, *Industry Week*, November 3, 1997, 18.

Stewart, C.T., Allocation of resources to health, *Journal of Resources to Health*, 6, 1971.

Stewart, C.T., *Healthy, Wealthy, or Wise? Issues in American Health Care Policy*, M.E. Sharpe, Armonk, NY, 1995.

Subcommittee on Science, Research, and Development, *The International Biological Program, Its Meaning and Needs*, U.S. House of Representatives, Washington, D.C., 1968.

Sutton, P., A speculative sketch of a successful ecologically-sustainable economy, Green Innovations Inc. 1998, Website http://www.peg.apc.org/~psutton/sectors.htm, 5/12/99.

Swift, A., *Global Political Ecology*, Pluto Press, London, 1993.

Tainter, J.A., *The Collapse of Complex Societies*, Cambridge University Press, Cambridge, 1988.

Toulmin, S., *Cosmopolis: The Hidden Agenda of Modernity*, University of Chicago Press, Chicago, 1991.

Tripp, R. and van der Heide, W., The erosion of crop genetic diversity and uncertainties, *Natural Resource Perspectives*, March 7, 1996, Overseas Development Institute, London, 1996.

Ulanowicz, R., *Ecology, The Ascendant Perspective*, Columbia University Press, New York, 1997.

United Nations Conference on Environment and Development, Rio de Janeiro, June 1992, Agenda 21, http://www.igc.apc.org/habitat/agenda21/ 7/29/99.

United Nations Development Program, *United Nations Human Development Report 1998*, Oxford University Press, New York, 1998.

U.S. Department of Agriculture, Agriculture FACT BOOK 98, http://www.usda.gov/ news/pubs/fbook98/chart1.htm, 3/18/99.

U.S. Department of Agriculture, *Agricultural Resources and Environmental Indicators, 1996-97*, Economic Research Service, Natural Resources and Environment Division Agricultural Handbook No. 712., Washington, D.C., July 1997.

U.S. Department of Agriculture, Questions and Answers about the National Organic Program Proposed Rule, at http://www.ams.usda.gov/nop/rule/Resource %20Material%20Files/qs&as.htm. 5/18/99.

U.S. Department of Agriculture Economic Research Service, Change in Livestock Production, 1969–1992, AER-754, August 1997 at http://www.econ.ag.gov/ epubs/htmlsum/are754.htm, 5/11/99.

U.S. Department of Health and Human Services, Clinton Administration Record on HIV/AIDS, Fact Sheet, March 12, 1996, http://www.hhs.gov/news/press/ 1998pres/980312b.html. 3/9/99.

U.S. Department of Transportation, *Transportation Statistics 1994*, U.S. Department of Transportation, Bureau of Transportation Statistics, Washington, D.C., 1994.

U.S. Environmental Protection Agency, Extended Product Responsibility: A New Principle for Product-Oriented Pollution Prevention, June 1997, http://www.epa.gov/epaoswer/non-hw/reduce/epr/epr.htm, 4/6/99.

Usui, M., National and international policy instruments and institutions for eco-restructuring, in *Eco-restructuring: Implications for Sustainable Development*, Ayres, R.U. (Ed.), United Nations University Press, Tokyo, 1998, 364.

van der Werf, E., Can ecological agriculture meet the Indian farmer's needs? in *Let Farmers Judge*, Hiemstra, W., Reijntjes, C., and van der Werf, E. (Eds.), Intermediate Technology Publications, London, 1992, 175.

Van Dieren, W., *Taking Nature into Account: A Report to the Club of Rome*, Springer-Verlag, New York, 1995.

Van Valen, L., Body size and numbers of plants and animals, *Evolution*, 27, 27, 1973.

Vitousek, P.M., Ehrlich, P.R., Ehrlich, A.H., and Matson, P.A., Human appropriation of the products of photosynthesis, *Bioscience*, 36(6), 368, 1986.

Vitousek, P.M., Mooney, H., Lubchenko, J., and Melillo, J., Human domination of Earth's ecosystems, *Science*, 277(25), 494, July 1997.

Von Neumann, J. and Morgenstern, O., *Theory of Games and Economic Behavior*, Princeton University Press, Princeton, NJ, 1953.

Von Weizsacker, E., Ulrich, E., Lovins, A.B., and Lovins, L.H., *Factor Four: Doubling Wealth — Halving Resource Use*, Earthscan, London, 1997.

Wachtel, P., *The Poverty of Affluence: A Psychological Portrait of the American Way of Life*, Free Press, New York, 1983.

Wackernagel, M. and Rees, W.E., *Our Ecological Footprint: Reducing Human Impact on the Earth*, New Society Publishers, Gabriola Island, B.C., Canada, 1996.

Wackernagel, M., Onisto, L., Callejas Linares, A., Lopez Falfan, I.S., Mendez Garcia, J., Suarez Guerrero, A.I., Suarez Guerrero, Ma. G., *Ecological Footprints of Nations, Report of the Centre for Sustainability Studies*, Universidad Anahuac de Xalapa, Mexico and The Earth Council, Costa Rica, 1997.

Wackernagel, M., Can trade promote an ecologically secure world? The global economy from an ecological footprint perspective, *Buffalo Environmental Law Journal*, 5(2), 180, 1998.

Walliman, I., Downsizing industrial society: ways to distribute scarcity and enhance the ability to sustain life, in *An Aging Population, an Aging Planet and a Sustainable Future*, Ingman, S.R. (Ed.), Proceedings from the conference held at the University of North Texas, February 26–28, 1995. Texas Institute for Research and Education on Aging, University of North Texas, Center for Texas Studies, Denton, TX, 1995.

Walzer, M., *Spheres of Justice: A Defense of Pluralism and Equality*, Basic Books, New York, 1983.

Wapner, P.K., *Environmental Activism and World Civic Politics*, State University of New York Press, Albany, 1996.

Warick, J., Cultivating farms to soak up a greenhouse gas, *Washington Post*, A03, November 23, 1998.

Waring, M., *If Women Counted, A New Feminist Economics*, HarperCollins, New York, 1988.

Weatherford, J.M., *Indian Givers: How the Indians of the Americas Transformed the World*, 1st ed., Crown Publishers, New York, 1988.

Weber, M., *The Protestant Ethic and the Spirit of Capitalism*, Charles Scribner's Sons, New York, 1958.

Weisman, A., *Gaviotas: A Village to Reinvent the World*, Chelsea Green, White River Junction, VT, 1998.

Westra, L. and Werhane, P., (Eds.), *The Business of Consumption: Environmental Ethics and the Global Economy*, Rowman and Littlefield, Lanham, MD, 1998.

Westra, L., Institutionalized violence and human rights, *Global Ecological Integrity*, Pimentel, D. (Ed.), Island Press, Washington, D.C., In Press.

Wilkinson, R.G., *Unhealthy Societies: The Afflictions of Inequality*, Routledge, New York, 1996.

Williamson, H., China's toy industry tinderbox, *Multinational Monitor*, 15(9), 24, 1994.

Witzeling, R., *People, Not Profit, the Story of the Credit Union Movement*, Kendall/Hunt, Dubuque, IA, 1993.

Wohlmeyer, H., Agro-eco-restructuring: potential for sustainability, in *Eco-restructuring: Implications for Sustainable Development*, Ayres, R.U. (Ed.), United Nations University Press, Tokyo, 1998, 276.

Wolff, E.N., Recent Trends in Wealth Ownership, a paper for the conference on *Benefits and Mechanisms for Spreading Asset Ownership in the United States*, New York University, December 10–12, 1998.

The Wonder Years, changes in franchising and the Franchise 500 since 1979, Entrepreneur, 27(1), 180, January 1999, http://web3.infotrac.galegroup.com/itw/session/526/624/11406033w5/, 4/15/99.

Wood, E., *Social Planning, A Primer for Urbanists*, Pratt Institute, Brooklyn, NY, 1965.

World Commission on Environment and Development, *Our Common Future*, Oxford University Press, New York, 1987.

World Health Organization, Micronutrient malnutrition — Half the world's population affected. *World Health Organization*, 78, 1, November 13, 1996.

World Resources Institute, *The Going Rate: What It Really Costs to Drive*, World Resources Institute, Washington, D.C., 1992.

Worldwatch Institute, *State of the World*, Worldwatch Institute, Washington, D.C., 1989.

Young, J. and Sachs, A., *The Next Efficiency Revolution: Creating a Sustainable Materials Economy, Worldwatch Paper 121*, Worldwatch Institute, Washington, D.C., 1994.

Zuger, A., Fever pitch: getting doctors to prescribe is big business, *New York Times*, A1, A13, January 11, 1999.

Index

A

Academia and science
  IBP experiment, 108–111
  products vs. processes, 76–78, 107
  selective R&D investment in, 107–108
Agenda 21, 7
Agriculture, 30
  commoditization through plant patents, 35–36
  Communist policies impact on, 209
  diversity benefits, 39–40, 84
  evolution of practices
    commoditization of inputs and outputs, 79–80
    farmer disadvantages in food economy, 80
    increase in energy use, 45
    products vs. processes, 78–79, 80–82
    R&D investments in, 94–96
  indigenous systems
    achievements of farmers, 84–86
    contribution to civilization, 83–84
    evolution of, 82–83
    loss of skills, 40
    loss of traditional knowledge, 86
  productivity and yield comparisons
    commoditized attributes, 88
    decline in skills and training, 87–88
    industrial vs. small farms, 86–87, 91
    low energy efficiency of modern, 89
    organic certification program, 93–94
    organic farming startup costs, 89
    organic foods as market niche, 91–92
    organic techniques, need for investment in, 93, 95
    organic vs. nonorganic research spending, 89–90

revolution in Britain, 162–163
standardization of, 39
subsidies and commoditization, 229–230
systems vs. product approach, 41–42
Allocation principles, 5
Alternative health care products, 98–99
American economy
  capitalistic system reform and adaptation, 169–170
  foundations of development, 167–169
Appropriate technology and ecosystem thinking
  ecosystem approach, 190–191, 192
  effective conservation attributes, 192–193
  extended producer responsibility, 193
  implication of, 188
  logic of commodity production, 191–192
  natural flow control consequences, 189–190
  natural flows study, 188–189
  natural flows vs. commoditization, 195
  shift in perspective needed, 193–194
  skills required, 188
  variability importance, 191
Automobiles, *see also* Transportation
  commoditized development, 106–107
  global per capita car ownership, 145

B

Barbie dolls, 25
Bills of exchange, 157–158
Bonfoc people, 133–134
Britain and commoditization economy
  agricultural revolution, 162–163
  legislative dominance effects, 154–155
  Parliamentary reforms, 154
  tax control, 153